Introduction to stellar astrophysics

Volume 2
*Stellar atmospheres*

Introduction to Stellar Astrophysics

Volume 1
*Basic stellar observations and data*
ISBN 0 521 34402 6 (hardback)
ISBN 0 521 34869 2 (paperback)

Volume 2
*Stellar atmospheres*
ISBN 0 521 34403 4 (hardback)
ISBN 0 521 34870 6 (paperback)

Volume 3
*Stellar structure and evolution*
ISBN 0 521 34404 2 (hardback)
ISBN 0 521 34871 4 (paperback)

# Introduction to stellar astrophysics

## Volume 2
*Stellar atmospheres*

Erika Böhm-Vitense
*University of Washington*

CAMBRIDGE
UNIVERSITY PRESS

Published by the Press Syndicate of the University of Cambridge
The Pitt Building, Trumpington Street, Cambridge CB2 1RP
40 West 20th Street, New York, NY 10011–4211
10 Stamford Road, Melbourne, Victoria 3166, Australia

First published 1989
Reprinted 1993 (with corrections)

Printed in Great Britain at the University Press, Cambridge

*British Library cataloguing in publication data*

Böhm-Vitense, Erika, 1923–
Introduction to stellar astrophysics.
Vol. 2: Stellar atmospheres
1. Stars
I. Title
523.8

*Library of Congress cataloguing in publication data*

Böhm-Vitense, E.
Introduction to stellar astrophysics.
Includes index.
Contents: v. 1. Basic stellar observations and data –
v. 2. Stellar atmospheres.
1. Stars. 2. Astrophysics.   I. Title.
QB801.B64 1989
523.8 88-20310

ISBN 0 521 34403 4 hardback
ISBN 0 521 34870 6 paperback

MC

# Contents

# Preface

In Volume 2 of *Introduction to Stellar Astrophysics* we will deal mainly with stellar atmospheres. What are stellar atmospheres? We have seen in Volume 1 that stars have temperatures starting at about 3000 K for the coolest stars up to somewhere around 40 000 K for the hottest stars. With such high temperatures stars certainly cannot be solid; they must all be in a gaseous phase. Therefore, the atmosphere cannot be defined as a gaseous layer on top of a solid core as on the Earth; there are no solid cores in the stars. Instead, astronomers define the atmosphere as those layers of the star from which we get the radiation. This means, of course, that this is the layer of the star about which we can obtain direct information. We see no photons from beneath the layer we call the atmosphere. All the radiation which originally came from deeper layers has been absorbed once or many times by atoms in the overlying layers and is finally emitted by an atom in the stellar atmosphere. The photons we receive tell us directly only about the condition of the atoms from which they were last emitted and those are the atoms in the stellar atmosphere. This is why we devote all of Volume 2 to stellar atmospheres.

How thick is this stellar atmosphere? When we discussed absorption in the Earth's atmosphere in Volume 1, we saw that the intensity of a light beam passing through a gas is diminished by a factor $e^{-\tau_\lambda}$, where $\tau_\lambda$ is the so-called optical depth of the layer of gas along the beam of light. This means that along a path with optical depth $\tau_\lambda = 1$ about two-thirds of the light is absorbed, and only one-third of the photons remain. Hence very few photons come from layers with an optical depth which is larger than 1. In a very crude way therefore we can say that the atmosphere of a star is the layer for which the optical depth is about 1. How thick is this layer actually in cm or km? We do know that the optical depth is defined as

$$\tau_\lambda = \int_0^s \kappa_\lambda \, ds,$$

where the integral is taken along the path $s$ of the light beam. $\kappa_\lambda$ is the absorption coefficient of the gas through which the beam passes. We first have to determine the nature of this gas, and then we have to measure the absorption coefficient of a similar gas in the laboratory to determine $\kappa_\lambda$ or we have to calculate it from atomic theory. Without further preparation at this point we can only say that the $\kappa_\lambda$ for stellar atmospheres have been determined (we will discuss how later) and we find that the atmosphere of the sun is roughly 100 km thick. For the hottest main-sequence stars it is about 1000 km thick. This thickness has to be compared to the radius of the sun, for instance, which is 700 000 km. The atmosphere is only a *very* thin surface layer. If we compare the thickness of the skin of an apple with the size of the apple, then this skin is relatively thicker than the atmosphere compared to the size of the star. It is somewhat disappointing that this is all we can see of the stars – such a very thin surface layer – yet we have to get all our information about the stars from the study of the radiation coming from this very thin surface layer. How much does this tell us about the interior of the star, about those very deep layers which we cannot see? This we will find out in Volume 3.

From the discussion above, it appears clear that we have to make a very sophisticated study of the starlight in order to obtain all the information we possibly can. This will involve quite a bit of theory. I will do my best to make this theory as transparent as possible. In order to do this I have to make some simplifications, but still retain the important ingredients. I am sure that some knowledgeable astrophysicists will object to some of the simplifications, but my experience has taught me that it is more important for the beginner to understand the basic principles, even at the expense of accuracy, than it is to see all the complications immediately. Once the basic physics is understood, the complications can always be added later.

I am also sure that several readers will think that some of the discussions are too elementary. Again, my experience has shown that many students have never heard about these elementary facts about atomic physics and the formation of spectral lines. In fact, I have talked to many well-known physicists who did not know that stellar spectra show absorption lines and not emission lines, as do laboratory spectra. I therefore feel that these basic facts need explanation, so that some readers are not lost from the very beginning.

Before we start our theoretical discussions, we review briefly those sections of Volume 1 which are important for the following discussion. A few new considerations will be inserted. The reader who has studied Volume 1 is advised only to skim through the first two chapters of Volume 2.

I apologize for not giving all the references for quoted investigations. Instead, I have chosen to give only references for other textbooks or reviews in which these references can be found. In addition, I give references for figures and tables which were reproduced from other books or papers. I also give references of recent articles, for which the references cannot be found in older textbooks.

In many sections this text relies heavily on the book *Physik der Sternatmosphären* by A. Unsöld, my highly respected teacher to whom I am very grateful.

I am very much indebted to Ms Sandi Larsen for typing this manuscript and for her patience with many changes.

# 1

# Stellar magnitudes and stellar colors

## 1.1    The apparent magnitudes

The easiest quantity to measure for a star is its brightness. It can be measured either by intercomparison of the brightness of different stars or by a quantitative measurement of the energy received on Earth. For a bright enough star – the sun, for instance – the latter could be done in a fundamental but not very accurate way by having the sun shine on a well-insulated bowl of water for a certain length of time and then measuring the increase in temperature. Knowing how much water is in the bowl and knowing the surface area of the water, we can calculate the amount of heat energy received from the sun per $cm^2$ s. The amount received will, of course, depend on the direction of the light beam from the sun with respect to the plane of the water surface (see Fig. 1.1). If the angle between the normal to the water surface and the direction of the beam is $\delta$ (see Fig. 1.1a), then the effective cross-section of the beam $t$ is smaller than the actual surface area of the water by a factor of $\cos \delta$. In Fig. 1.1a, the amount of heat energy received per $cm^2$ s will therefore be smaller by this factor than in Fig. 1.1b, where the sunlight falls perpendicularly onto the bowl of water.

For stars this method in general will not work; we would have to wait too long before we could measure an increase in temperature, and it is very difficult to prevent the water surface from cooling again in the meantime. We need much more sensitive instruments to measure radiation. We have to calibrate photomultipliers or photon-counting devices against a standard light source for which the amount of energy emitted per $cm^2$ s is known. Such standard light sources are black bodies (see Volume 1, Chapter 4, and Chapter 3 in Volume 2). Another standard source is the cyclotron radiation in large particle accelerators.

Relative brightnesses of stars had been measured long before these modern instruments became available, but even now it is very difficult to measure accurate, absolute fluxes received from the stars because of absorption in the Earth's atmosphere.

If we measure only relative brightnesses of stars, i.e., if we only *compare* brightnesses of stars, then things become much easier. In principle we need only *one standard star* to which we can compare all the others. However, it might happen that the chosen standard star turned out to be variable, and then we would be in trouble. Therefore, a large number of stars was chosen – the so-called north polar sequence, or, more recently, the standard stars studied by Morgan and Johnson. For all practical purposes, however, our one standard star is *Vega*, also called α Lyrae, i.e., the brightest star in the constellation Lyra.

The oldest observations were, of course, made by looking at the stars. The old astronomers called the brightest stars first magnitude, the next brightest stars second magnitude, etc. This magnitude system is still in use today, although it has been put on a quantitative basis. It still causes much confusion for students not yet accustomed to it because the magnitudes become *smaller* when the stars are *brighter*.

Since the old observations were made by visual comparison of stars, the magnitude scale is determined by the sensitivity of our eyes, which is logarithmic. This means a given brightness difference as sensed by our eyes corresponds to a given factor for the intensities. The magnitude scales are therefore logarithmic scales.

Let us call the amount of energy $E$ received from a star per cm² s perpendicular to the beam of light

$$E = \pi f,\qquad\qquad 1.1a$$

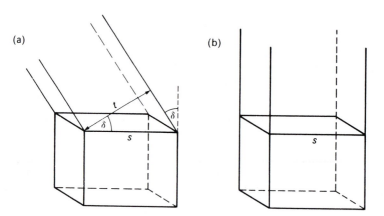

*Fig. 1.1.* The amount of energy received to heat the water in the bowl depends on the angle at which the sunbeams enter the bowl. If the beam is inclined with respect to the normal on the water surface $s$ by an angle $\delta$, the effective beam width is reduced by a factor $\cos \delta = t/s$.

or, if we refer to a given wavelength $\lambda$ and a bandwidth of 1 cm

$$E_\lambda = \pi f_\lambda, \qquad 1.1b$$

where $f_\lambda$ is called the *apparent flux*.

The magnitudes referring to the energy received here on Earth (that is, to the brightness as it appears to us) are called *apparent magnitudes*. They are designated by a *lower case m*, or $m_\lambda$ if they refer to a given wavelength.

These magnitudes are related to the energy received by

$$m(1) - m(2) = -2.5 \log \frac{\pi f(1)}{\pi f(2)}, \qquad 1.2$$

where (1) refers to star 1 and (2) refers to star 2.

If we refer to a given wavelength band (for instance, to visual observations), then we have

$$m_V(1) - m_V(2) = -2.5 \log \frac{\pi f_V(1)}{\pi f_V(2)}. \qquad 1.3$$

Obviously, this does not determine the magnitudes of the stars unless we give the magnitude for one star. This star is Vega; **it always has the magnitude 0 by definition, regardless of the wavelength we are looking at.\*** The only exception is the *bolometric magnitude*, which refers to the total radiation of the stars, including the invisible part.

**The apparent magnitudes refer to the energies received above the Earth's atmosphere,** i.e., after correction for the absorption occurring in the Earth's atmosphere. In the next section we will see how we can make this correction.

We want to emphasize here that this definition of the zero point for all magnitude scales is quite arbitrary and has nothing to do with the actual energy distribution of Vega. It has its foundation only in history. When the magnitude measurements were made, one could only observe first with the eye, then later with photographic plates, and it was not known how the energies in these wavelength bands were related. It therefore seemed best to keep the same normalization.

In Fig. 1.2 we show the actual energy distribution of Vega as a function of wavelength $\lambda$, i.e., $f_\lambda$ is plotted as a function of $\lambda$. It is obvious that $f_\lambda$ is not independent of $\lambda$, even though all magnitudes $m_\lambda = 0$ by definition. It is very important to remember this if we use the magnitudes to infer anything about the energy distribution in the stars.

## 1.2 Stellar colors

The brightness of a star in general depends on the wavelength at which the star is observed. From Fig. 1.2 we see, for instance, that Vega

---

\* Actually the average of stars like Vega. For Vega $m_\lambda$ may be 0.01 or 0.02.

emits much more energy at the shorter wavelengths than at the longer wavelengths. If we observe a star which is fainter than Vega in the blue region of the spectrum, but is just as bright as Vega in the visual region, then this star has a visual magnitude equal to that of Vega, namely $m_V = 0$, but in the blue it has a larger magnitude than Vega because it is fainter. This may well happen if we compare a star like the sun with Vega. In Fig. 1.3 we compare the energy distribution of a solar type star with that of Vega. We have assumed the solar type star to be at such a distance that the observed fluxes in the visual spectral region are the same. In the blue and in the ultraviolet the solar type star is then fainter than Vega. For the solar type star the blue magnitude is larger than the visual magnitude. For a given star the magnitudes in different wavelength bands can be quite different. The difference in these magnitudes tells us something about the energy distribution in the stars. The solar type star has relatively more energy in the visual region than Vega; it is more red than Vega. If we take the difference between the magnitude measured in the blue, $m_B$, and in the visual, $m_V$, we find for the solar type stars

$$m_B - m_V > 0. \tag{1.4}$$

For the sun $m_B - m_V = 0.63$. This difference $m_B - m_V$ is usually abbreviated as

$$m_B - m_V = B - V. \tag{1.5}$$

**For stars redder than Vega we then find B − V > 0.**

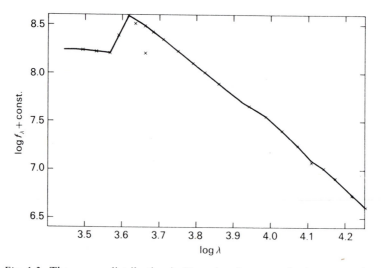

Fig. 1.2. The energy distribution in Vega. $\log f_\lambda + $ const. is shown as a function of wavelength $\lambda$. $f_\lambda$ is strongly variable with wavelength, even though all $m_\lambda$ are zero by definition.

According to our discussions in Volume 1, such stars are generally expected to be cooler than Vega. Stars with relatively more energy in the blue than Vega have smaller magnitudes in the blue; they therefore have $B - V < 0$. These stars are expected to be hotter than Vega.

We can, of course, determine magnitudes for many wavelength bands and correspondingly determine many colors. Different astronomers prefer different color systems, depending on what they want to infer from these colors. There are, therefore, many different color systems in use. The most widely used system is still the UBV system, using ultraviolet, blue and visual magnitudes. The bandwidths for these magnitudes are roughly 1000 Å, which means they are broad bands. The sensitivity curves for this system are reproduced in Fig. 1.4. Of course, we receive more light in such broad bands than in narrow bands, which is an advantage if we want to observe faint stars. On the other hand, much information is smoothed out

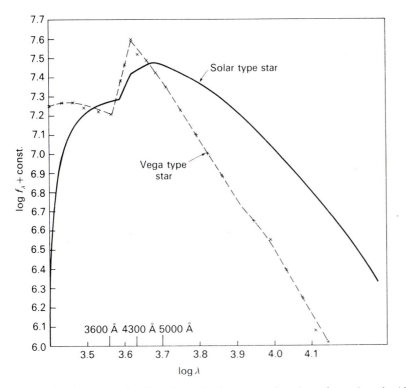

*Fig. 1.3.* $\log f_\lambda + \text{const.}$ for Vega is again shown as a function of wavelength. Also shown is the relative energy distribution of a star like the sun. This star is assumed to be at a distance such that its brightness in the visual is the same as that of Vega. We see that it then has much less flux than Vega in the blue wavelength region around 4300 Å. Its $m_B$ must then be larger than zero. For this star $B - V \sim 0.6 > 0$.

in such broad bands. For brighter stars narrower bands give us more information. We will discuss other color systems later, such as Strömgren's color system.

## 1.3    Correction for the absorption in the Earth's atmosphere

We discussed in Volume 1 how we can determine the absorption of solar light in the Earth's atmosphere, also called extinction, by measuring the radiation received on Earth for different zenith distances of the sun. For stars we can follow a similar procedure (see Fig. 1.5). When light passes through the Earth's atmosphere the intensity in the beam is changed slightly when the beam proceeds along the path element d$s$. The more

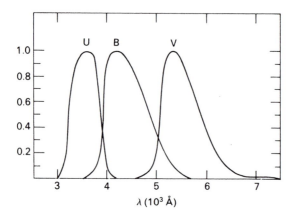

Fig. 1.4. Sensitivity curves for the receiving instruments to measure UBV magnitudes. (From Johnson, 1965.)

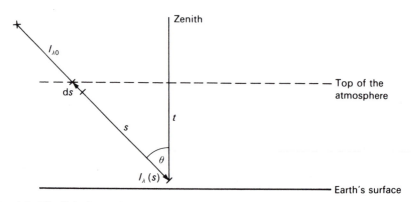

Fig. 1.5. The light beam from the star passes through the Earth's atmosphere along the path $s$, which is inclined with respect to the zenith direction by the angle $\theta$. Along path $s$ the intensity is reduced due to absorption. The absorption increases if $s$ becomes longer for larger zenith distances.

energy there is in the beam, i.e., the more photons there are, the greater the chance that one will be absorbed. The change in energy $dI_\lambda$ is therefore proportional to $I_\lambda$, where $I_\lambda$ is the intensity of the beam at wavelength $\lambda$. The number of photons absorbed at the wavelength $\lambda$ also depends on the special properties of the gas in the atmosphere, which is described by the so-called *absorption coefficient* $\kappa_\lambda$. The change in intensity along the path length $ds$ is then given by

$$dI_\lambda = -\kappa_\lambda I_\lambda \, ds. \qquad 1.6$$

Since the intensity is decreasing, the $dI_\lambda$ is, of course, negative. Dividing by $I_\lambda$ and remembering that $dI/I = d \ln I$, where $\ln$ denotes the logarithm to the base e, we find

$$d(\ln I_\lambda) = -\kappa_\lambda \, ds = -d\tau_\lambda. \qquad 1.7$$

Here we have defined the so-called *optical depth* $\tau_\lambda$ by

$$d\tau_\lambda = \kappa_\lambda \, ds \quad \text{and} \quad \tau_\lambda(s_0) = \int_0^{s_0} \kappa_\lambda \, ds. \qquad 1.8$$

Equation (1.8) can be integrated on both sides and yields

$$\Delta(\ln I_\lambda) = \ln I_\lambda(s) - \ln I_\lambda(0) = -\int_0^s \kappa_\lambda \, ds = -\int_0^{\tau_\lambda(s)} d\tau_\lambda = -\tau_{\lambda s}(s), \quad 1.9$$

where $\tau_{\lambda s}(s)$ is the optical depth along the path $s$.

With $\ln I_\lambda(s) - \ln I_\lambda(0) = \ln \dfrac{I_\lambda(s)}{I_\lambda(0)}$ we get

$$\ln \frac{I_\lambda(s)}{I_\lambda(0)} = -\tau_{\lambda s}(s). \qquad 1.10$$

Taking the exponential on both sides gives

$$I_\lambda(s) = I_\lambda(0)e^{-\tau_{\lambda s}(s)}. \qquad 1.11$$

The optical depth along the path of the light $\tau_{\lambda s}$ depends on the zenith distance $\theta$, as can be seen in Fig. 1.5. We see that $\cos\theta = t/s = dt/ds$ or

$$ds = \frac{dt}{\cos\theta} = \sec\theta \, dt. \qquad 1.12$$

It then follows that

$$\tau_{\lambda s} = \int_0^s \kappa_\lambda \, ds = \sec\theta \int_0^t \kappa_\lambda \, dt = \sec\theta \tau_{\lambda t}, \qquad 1.13$$

where $\tau_{\lambda t}$ is the optical depth measured perpendicularly through the atmosphere. We can then write equation (1.10) in the form

$$I_\lambda(s, \theta) = I(0)e^{-\sec\theta \tau_{\lambda t}}, \qquad 1.14$$

where $\tau_{\lambda t}$ is now independent of $\theta$. $\tau_{\lambda t}$ is called the optical depth of the atmosphere at wavelength $\lambda$ and is usually written as $\tau_\lambda$.

In order to derive from $I_\lambda(s, \theta)$ the intensity above the Earth's atmosphere, $I_\lambda(0)$, we have to know $\tau_\lambda$, which may depend strongly on wavelength.

Just as we discussed for the sun in Volume 1, we can determine the optical depth $\tau_\lambda$ by measuring the intensity received from a given star by measuring $I_\lambda(s)$ for different angles $\theta$, when the star moves to different places during the course of the night (due to the Earth's rotation). In principle, two measurements are enough; they provide us with two equations for the two unknowns $I_\lambda(0)$ and $\tau_\lambda$, as we will see. Let $I_{\lambda,1} = I_\lambda(\theta_1, s)$ be the intensity measured for the zenith distance $\theta_1$ of the star and $I_{\lambda,2}$ the intensity measured at zenith distance $\theta_2$. We then find, according to equation (1.10),

$$\ln I_{\lambda,1} - \ln I_{\lambda,2} = -\tau_\lambda(\sec \theta_1 - \sec \theta_2), \qquad 1.15$$

and

$$\tau_\lambda = \frac{\ln I_{\lambda,1} - \ln I_{\lambda,2}}{(\sec \theta_2 - \sec \theta_1)}. \qquad 1.16$$

Because there are always measuring errors, it is safer to do many measurements during the course of the night and plot them in a diagram, as shown in Fig. 1.6. The best fitting line gives the relation

$$\ln I_\lambda = \ln I_\lambda(0) - \sec \theta \tau_\lambda.$$

The gradient of this line is determined by $\tau_\lambda$, and $\ln I_\lambda(0)$ can be read from the intersection of this line with the $\ln I_\lambda$ axis for $\sec \theta = 0$. (Never mind

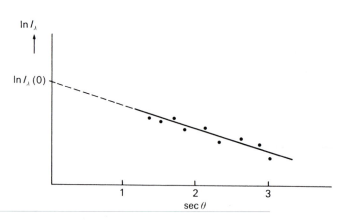

Fig. 1.6. In order to determine $I_\lambda(0)$ it is best to plot the measured values of $\ln I_\lambda(\theta)$ as a function of $\sec \theta$. The best-fitting straight line through the points intersects the $\ln I_\lambda$ axis at the value $\ln I_\lambda(0)$.

that $\sec \theta = 0$ does not exist; this is just a convenient way to read off the value for $\ln I_\lambda(0)$.)

In fact, Fig. 1.5 is a simplification of the geometry. The Earth's surface is not plane parallel. In addition, the beam of light is bent due to refraction. These effects give a relation $\tau_{\lambda s}/\tau_{\lambda t} \neq \sec \theta$. The actual ratio is called *air mass*. For $\sec \theta < 2$ the difference is in the third decimal place and is in most cases negligible.

It is important to note that the above derivation only holds for each $\lambda$ with a given $\kappa_\lambda$, or for a broad band if $\kappa_\lambda$ happens to be wavelength independent. It cannot be applied to a broad wavelength band with varying $\kappa_\lambda$. In the ultraviolet the variations of $\kappa_\lambda$ with $\lambda$ are especially strong.

People tend to forget the variations of $\kappa_\lambda$ when determining the extinction correction for broad band colors like the UBV magnitudes. Using the same method yields wrong values for $\tau_\lambda$.

How can we avoid this problem? We have to remember that ultimately we want to determine the ratio of the corrected intensity of one star to the corrected intensity of a standard star. If we correct both stars by the same wrong factor, the wrong factor does not do any harm because it cancels out. According to equation (1.14), the correction factor is $e^{\sec \theta \tau_\lambda}$. If we measure both stars for the same zenith distance, $\theta$, the errors will almost cancel out. (They do not completely cancel if the stars have different energy distributions, because then different errors in $\tau_\lambda$ will occur.) In addition, errors will be minimized if the stars are measured at small zenith distances.

In Table 1.1 we list the relation between $U - B$ and $B - V$ colors for normal main sequence stars and for supergiants. In the next section we will briefly review what these stars are. In the same table we also list the UBV colors for black bodies of different temperatures. As can be seen, the stars do not radiate like black bodies. One of the aims of Volume 2 is to understand why the stars have a different energy distribution from that of black bodies. As can be seen from Table 1.1, the stars have less energy in the ultraviolet.

In Fig. 1.7 we have plotted the $U - B$ colors as a function of $B - V$ for stars and for black bodies.

## 1.4    Absolute magnitudes of stars

For the stellar astrophysicist it is more important to know the amount of radiation leaving the star than it is to know how much we receive; the latter being largely determined by the distance of the star. In order to compare the intrinsic brightnesses of the stars we either have to

Table 1.1.

| B − V | U − B Main sequence | U − B Supergiants | U − B Black body |
|---|---|---|---|
| − 0.30 | − 1.08 | − 1.16 | − 1.22 |
| − 0.20 | − 0.71 | − 0.94 | − 1.11 |
| − 0.10 | − 0.32 | − 0.65 | − 1.00 |
| 0.00 | 0.00 | − 0.34 | − 0.89 |
| + 0.10 | + 0.10 | − 0.02 | − 0.78 |
| + 0.20 | + 0.11 | + 0.23 | − 0.68 |
| + 0.30 | + 0.07 | + 0.31 | − 0.57 |
| + 0.40 | + 0.01 | + 0.34 | − 0.46 |
| + 0.50 | + 0.03 | + 0.38 | − 0.35 |
| + 0.60 | + 0.13 | + 0.41 | − 0.24 |
| + 0.70 | + 0.26 | + 0.47 | − 0.13 |
| + 0.80 | + 0.43 | + 0.55 | − 0.00 |
| + 0.90 | + 0.63 | + 0.66 | + 0.11 |
| + 1.00 | + 0.81 | + 0.81 | + 0.23 |
| + 1.10 | + 0.96 | + 0.96 | + 0.35 |
| + 1.20 | + 1.10 | + 1.10 | + 0.48 |
| + 1.30 | + 1.22 | + 1.22 | + 0.61 |

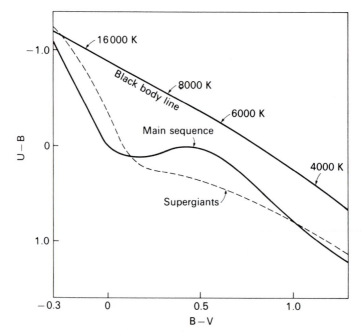

*Fig. 1.7.* The two-color diagram for the U − B, B − V colors for stars and for black bodies.

determine the distances of the stars or we have to compare stars which happen to be at the same distance, namely stars in star clusters. We saw in Volume 1 how we can determine distances for nearby stars by means of trigonometric parallaxes. If for these nearby stars we determine the magnitudes they would have if they were at a distance of 10 pc (i.e., if we determine $m$ (10 pc) by comparison with the actual brightness of Vega which is $m = 0$), then this **$m$ (10 pc) is called the** *absolute magnitude* **of the star**. Since Vega is closer than 10 pc, its absolute magnitude is larger than 0; it would be fainter than it actually is if it were at 10 pc. Its absolute magnitude is $M_V = 0.5$. Here we have used a capital $M$ to indicate that we are talking about absolute magnitudes. **Absolute magnitudes are always given by capital $M$**; the index refers again to the wavelength band used to determine the absolute magnitude. As was derived in Volume 1, the absolute magnitude can be obtained from the apparent magnitude and the distance $d$ (in parsecs) of the star as follows:

$$M_V - m_V = 5 - 5 \log d. \qquad 1.17$$

The abbreviation log means logarithm to the base 10. Similar relations hold, of course, for the magnitudes in other wavelength bands. It is obvious that the colors determined from the absolute magnitudes are the same as the ones determined from the apparent magnitudes.

## 1.5 The luminosities of the stars

The total amount of radiation leaving the stellar surface per second is called the *luminosity* of the star. If $\pi F$ is the total amount of radiation leaving 1 cm$^2$ of the stellar surface per second going into all directions, then the luminosity $L$ of the star is

$$L = 4\pi R^2 \pi F. \qquad 1.18$$

Here $R$ is the radius of the star and $F$ is called the *surface flux*. Since the same amount of radiation must go through a sphere with radius $d$ some time later (provided there is no absorption on the way by, for instance, interstellar matter), the luminosity $L$ must also be given by

$$L = 4\pi d^2 \pi f. \qquad 1.19$$

Together, these two equations yield

$$f = F \times (R^2/d^2), \qquad 1.20$$

where $R/d$ is the *angular radius* of the star.

In Volume 1 we saw how angular radii can be determined for bright blue stars or for large stars. For the sun everybody can measure the angular radius – it turns out to be $\sim 1/200$. Surface fluxes can be determined rather easily, even if we do not know the distance of the star, provided we can measure its angular radius.

The ratio of the luminosities of two stars can also be described by means of magnitudes. The magnitudes which measure the luminosity are called *bolometric magnitudes* and they are abbreviated by $m_{bol}$ or $M_{bol}$. Actually, $M_{bol}$ measures the true luminosity of the star, but sometimes it is also convenient to use $m_{bol}$, as we shall see. These bolometric magnitudes measure all the radiation leaving the star. They also include the radiation from parts of the spectrum which we cannot see because it is either too far in the infrared, where the Earth's atmosphere absorbs the light, or the radiation may be in the ultraviolet, where the Earth's atmosphere also absorbs all the light. It is even worse for the extreme ultraviolet light, i.e., for wavelengths shorter than 912 Å. In this region of the spectrum the interstellar medium absorbs all the light for stars further away than about 50 parsecs. Even from satellites we have no chance of observing this radiation (unless we observe in the X-ray region where the interstellar medium becomes more transparent). We have to rely on theoretical extrapolations to estimate the amount of radiation emitted in this region of the spectrum. If we compare the visual and the bolometric magnitudes, it is clear that there is more energy coming out in the whole spectrum than just in the visual. The bolometric magnitudes are therefore generally smaller than the visual magnitudes, although there are some exceptions for the supergiants. This peculiar situation is due to the unfortunate calibration of the bolometric magnitudes. The difference between visual and bolometric magnitudes is called the *bolometric correction*, abbreviated to *BC*. This means

$$M_{bol} = M_V - BC. \qquad\qquad 1.21$$

Unfortunately, not all astronomers agree on the sign in equation (1.21). Often bolometric corrections are given as negative numbers if an author prefers a + sign in equation (1.21).

As we said before, the *BC* measures the amount of radiation not seen by the eye. This amount is large for red stars, where most of the radiation is in the infrared, and also very large for very hot stars, for which a large fraction of the radiation is in the ultraviolet. Between these extremes the *BC* has a minimum for stars with B − V colors around 0.3. Astronomers wanted to make the minimum *BC* on the main sequence equal to zero and normalized the bolometric magnitudes accordingly. The result is that the *BC* changes sign for supergiants of solar temperature. In Table 1.2 we list the bolometric corrections as a function of the B − V colors of the stars.

Table 1.2. *Bolometric corrections BC;* $M_{bol} = M_V - BC$

| B − V | Main sequence | | Supergiants | |
|---|---|---|---|---|
| | $T_{eff}$ | BC | $T_{eff}$ | BC |
| −0.25 | 24 500 | 2.30 | 26 000 | 2.20 |
| −0.23 | 21 000 | 2.15 | 23 500 | 2.05 |
| −0.20 | 17 700 | 1.80 | 19 100 | 1.72 |
| −0.15 | 14 000 | 1.20 | 14 500 | 1.12 |
| −0.10 | 11 800 | 0.61 | 12 700 | 0.53 |
| −0.05 | 10 500 | 0.33 | 11 000 | 0.14 |
| 0.00 | 9 480 | 0.15 | 9 800 | −0.01 |
| +0.10 | 8 530 | 0.04 | 8 500 | −0.09 |
| +0.2 | 7 910 | 0.00 | 7 440 | −0.10 |
| +0.3 | 7 450 | 0.00 | 6 800 | −0.10 |
| +0.4 | 6 800 | 0.00 | 6 370 | −0.09 |
| +0.5 | 6 310 | 0.03 | 6 020 | −0.07 |
| +0.6 | 5 910 | 0.07 | 5 800 | −0.03 |
| +0.7 | 5 540 | 0.12 | 5 460 | +0.03 |
| +0.8 | 5 330 | 0.19 | 5 200 | +0.10 |
| +0.9 | 5 090 | 0.28 | 4 980 | +0.19 |
| +1.0 | 4 840 | 0.40 | 4 770 | +0.30 |
| +1.2 | 4 350 | 0.75 | 4 400 | +0.59 |

## 1.6    The color magnitude diagram of the stars

Once we know the distances of the stars, we can determine their absolute magnitudes. It is quite instructive to plot the absolute magnitudes as a function of the stars' B − V colors, as we have done in Fig. 1.8 for the nearby stars for which trigonometric parallaxes can be measured fairly accurately. Most stars fall along one sequence, the so-called main sequence. There are a few stars which fall along an almost horizontal branch, these are called the giants. They are brighter than their main sequence counterparts with the same B − V. As we shall see, stars with a given B − V have about the same surface flux *F*. If indeed this is so, then, according to equation (1.18) the brighter stars must have larger radii; hence the name 'giants'. If we were to plot more distant stars in the same diagram, we would find stars which are even brighter than the giants. They must then be still larger and are called supergiants.

There are also a few stars below the main sequence, i.e., which are fainter. By the same kind of argument, they must then be smaller than the main sequence stars and we would expect them to be called dwarfs. However, this is not the case because the name 'dwarf' has been reserved for the main sequence stars. Since the first detected subluminous stars were rather bluish or white in color, they were called 'white dwarfs'. In the meantime, several of these stars have been found which are not bluish or white; there are also some rather red ones. So now we also have red 'white dwarfs'.

In Volume 3 we will understand why the stars are found along such sequences in the color magnitude diagram. Here it is only important to know what we mean by main sequence stars, giants, supergiants, and white dwarfs.

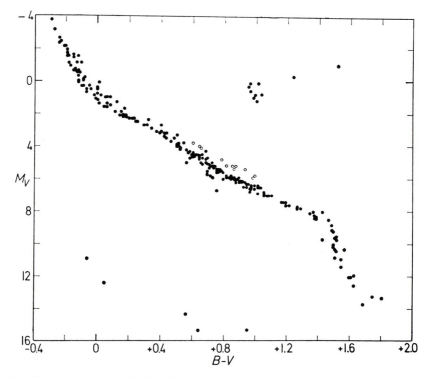

*Fig. 1.8.* The absolute magnitudes of nearby stars and some clusters are plotted as a function of the B − V colors. Most stars fall along a sequence, called the main sequence. Some stars are brighter than the main sequence stars; they are called giants. A few stars are much fainter than the main sequence stars; they are called white dwarfs. The open circles refer to probable binaries.

# 2

# Stellar spectra

## 2.1    The spectral sequence

If we look at the spectra of the stars we find that most spectra show a series of very strong lines which can now be identified as being due to absorption by hydrogen atoms. They are called the *Balmer lines*. Since these lines appeared to be strong in almost all stellar spectra, astronomers started to classify spectra according to the strengths of these lines, even though they did not know what caused them. The spectra with the strongest Balmer lines were called *A stars*, those with somewhat fainter lines were called *B stars*, and so on down the alphabet. It later turned out that within one such class there were spectra which otherwise looked very different; there were also stars within one spectral class which had very different colors. Taking this into account, the spectra were then rearranged mainly according to the $B - V$ colors of the stars which did, however, turn out to give a much better sequence of the spectra. Spectra in one class now really looked quite similar. Now the spectra with the strongest Balmer lines are in the middle of the new sequence. The bluest stars have spectral type O, the next class now has spectral type B, and then follow the A stars, the F stars, and then the G, K, and M stars. In Fig. 2.1 we again show the spectral sequence. The bluest stars – the O stars – are at the top. Their Balmer lines are rather weak, as the letter O indicates. The following B stars have much stronger Balmer lines; also other lines, for example those due to neutral helium, become stronger than in the O stars. In the A stars the Balmer lines are strongest, as indicated by the letter A. Further down the sequence we see an increasing number of other spectral lines which gain in intensity. These lines are mainly due to the absorption by heavy atoms like iron, chromium, titanium, etc. Astronomers usually call all the atoms with atomic weights larger than 5 the 'metals'. They are, of course, aware that not all these elements are metals; it is simply an abbreviation for the elements more massive than helium.

15

We will devote a large part of Volume 2 to the explanation and understanding of this spectral sequence, and to the analysis of the spectra in order to determine the chemical abundances in stars. In order to do so we will have to discuss the temperature and pressure stratification in the stars, depending on their spectral classes and luminosities.

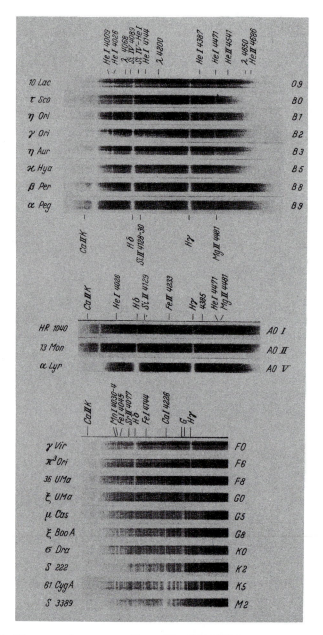

*Fig. 2.1.* The spectral sequence. (From Unsöld, 1977.)

It is customary among astronomers to call the blue stars *early type stars* and the red stars *late type stars*. This has nothing to do with the ages of the stars; it only means that the early type stars come first in the spectral sequence and the late type stars are at the end of the spectral sequence. The early type stars are blue and, as we shall see, hot and the late type stars are red and cool.

# 3

# Temperature estimates for stars

## 3.1    The black body

If we want to analyse the radiation of stars we must have laboratory light sources with which we can compare the radiation. For instance, if we want to determine the temperature of the stellar gas, we have to know how the radiation of a gas changes with temperature. We also need a light source whose radiation properties do not depend on the kind of material of which it is made, since *a priori* we do not know what the stars are made of. Such an ideal light source is the so-called *black body*.

What is a black body? We call something black if it does not reflect any light falling on it. In the absence of any radiation coming from the black body itself, it then looks black because no light falling on it is redirected or scattered into our eyes. If we want to determine temperatures from a comparison with an ideal light source, then this light source must have the same temperature everywhere. This means it must be in *thermodynamic equilibrium*, which means that it has reached a final state of equilibrium such that nothing will change in time. Such an ideal light source is best realized by a volume of gas inside a well-insulated box with a tiny hole in it. This hole is nearly a perfect black body because any light beam falling into this tiny hole will be reflected back and forth on the walls of this box (see Fig. 3.1) until it is finally absorbed either by the wall or by the gas in the box. The chances of the light getting out of this tiny hole again are extremely small. This does not mean that the hole cannot emit its own radiation, which we can measure when it comes out of the hole. Think about the windows in a house. If you look at an open window it looks black to you because the sunlight disappears in this hole; however, when the people in the room turn on the light the window will no longer be dark. Yet, the window still remains a fairly good black body because the sunlight will still disappear into the hole. The light is still all absorbed in the hole. The light which we see is emitted by the black body itself. But, of course, the room is not in thermodynamic equilibrium. If we look at the radiation from different parts in the room we see different radiation and different temperatures. The window is black, but it is not what we would call a black body to be used as an ideal light source. In our ideal light source the radiation in the box cannot depend on the properties of the gas or the walls. For instance, if one wall were to emit more radiation, say at 5000 Å, than the other walls, then there would be a flow of photons from this wall to the other walls, which would then be heated up. We could also use this flow of photons to power a *perpetuum mobile*. Yet, the second law of thermodynamics tells us that we cannot power a thermodynamic machine unless we have temperature differences such that a heat flux goes from the higher temperature region to the lower one. A black body in complete thermodynamic equilibrium by definition does not have temperature differences. If it has, it is not a black body. We have to insulate it better and wait longer until all temperature differences have disappeared. Then we find that the radiation coming out of the tiny hole always looks the same, no matter which gas is in the box nor what the box is made of. Obviously, the hole has to be tiny, otherwise the gas next to the hole will cool off and we will no longer have the same temperature everywhere. If we now measure the radiation leaving the hole we find, of course, that the amount getting out increases in proportion to the size of the hole. In order to get a measurement independent of size, we have to divide the amount of energy received by the area of the hole, and then determine the amount of radiation which we would receive if the hole had a cross-section of 1 cm$^2$. We also have to measure the amount of energy which we receive in each unit of time. The amount of energy which leaves the hole per cm$^2$ each

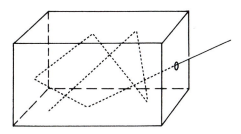

*Fig. 3.1.* The box with a small hole will act as a black body, i.e., will absorb all the light falling into the hole. If such a black body is well insulated and left to itself for a long time, complete thermodynamic equilibrium is established. The radiation of such a black body serves as an ideal light source with which stellar radiation can be compared.

second into all directions and at all wavelengths is called $\pi F$, where $F$ is called the *flux*. It is found that the flux increases with increasing temperature of the black body. Stefan and Boltzmann measured that

$$\pi F = \sigma T^4,\qquad\qquad 3.1$$

where $\sigma$ is the Stefan–Boltzmann constant,

$$\sigma = 5.67 \times 10^{-5} \text{ erg cm}^{-2} \text{ degree}^{-4} \text{ s}^{-1},$$

provided we measure the temperature $T$ in degrees Kelvin.

Equation (3.1) is called the *Stefan–Boltzmann law*.

If the temperature is high the hole no longer looks black to us because of its own radiation. Nevertheless, it is still a black body because it still absorbs all the light falling into the hole. The experiment verified that all black bodies, no matter what they are made of, emit the same kind of radiation if they have the same temperature. We can then once forever measure the energy distribution of a black body with a given temperature.

If we do not measure the total radiation coming out of the hole, but only the amount which comes out per second perpendicularly to the hole and into a cone of opening $d\omega$ (see Fig. 3.2) and in a wavelength band of width $d\lambda$, then the amount of energy is given by

$$E_\lambda = I_\lambda \, d\lambda \, d\omega \, \Delta\sigma.\qquad\qquad 3.2$$

where $\Delta\sigma$ is an area perpendicular to the beam of radiation.

$I_\lambda$ is called the intensity. For a black body the intensity distribution is given by the well-known *Planck formula*:

$$I_\lambda = \frac{2hc^2}{\lambda^5} \frac{1}{e^{hc/\lambda kT} - 1} = B_\lambda,\qquad\qquad 3.3$$

$c = $ velocity of light $= 3 \times 10^{10}$ cm,

$h = $ Planck's constant $= 6.62 \times 10^{-27}$ erg s,

$k = $ Boltzmann constant $= 1.38 \times 10^{-16}$ erg degree$^{-1}$.

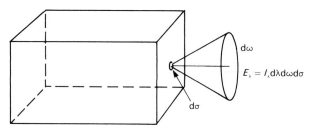

*Fig. 3.2.* The energy leaving the black body through a hole with an area $d\sigma$ into a cone with opening $d\omega$ and a band width $d\lambda$ is given by $E_\lambda = I_\lambda \, d\lambda \, d\omega \, d\sigma$. For $\Delta\sigma = 1 \text{ cm}^2$, $\Delta\lambda = 1$ cm and $\Delta\omega = 1$ this energy is equal to the intensity $I_\lambda$, which for a black body is called $B_\lambda$.

This is the amount of energy measured per wavelength interval $\Delta\lambda = 1$. If the energy per unit frequency interval is measured, we find

$$I_v = \frac{2hv^3}{c^2} \frac{1}{e^{hv/kT} - 1} = B_v = \text{black body intensity.} \qquad 3.4$$

Clearly, a frequency interval $\Delta\lambda = 1$ corresponds to a different wavelength band than the interval $\Delta v = 1$. For a given $\Delta\lambda$ we have $\Delta v = c/\lambda^2 \times \Delta\lambda$; the frequency interval $\Delta v = 1$ corresponds to a wavelength interval $\Delta\lambda = \lambda^2/c$, which is not equal to 1. Therefore, $I_v$ is 'larger' by the factor $\lambda^2/c$ than $I_\lambda$.

$B_\lambda$ and $B_v$ are common abbreviations for the black body energy distribution.

In Fig. 3.3 we show the logarithm (to the base 10) of $B_\lambda$ as a function of the wavelength $\lambda$ measured in angstroms (1 Å $= 10^{-8}$ cm). For increasing temperatures the intensity increases for all wavelengths. The shorter the wavelengths, the steeper the increase percentage-wise. The maximum in the energy distribution shifts to shorter wavelengths for higher temperatures. This is expressed quantitatively by *Wien's displacement law*:

$$\lambda_{\text{max}} T = \text{constant} = 0.289\,73 \text{ cm deg,} \qquad 3.5$$

where $\lambda_{\text{max}}$ is the wavelength at which we find the maximum for $I_\lambda$.

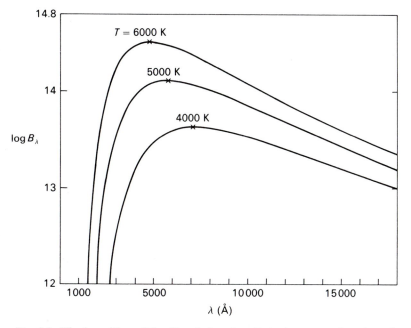

*Fig. 3.3.* The logarithm of the Planck function $B_\lambda$ is shown as a function of wavelength $\lambda$ for different temperatures of the black body. With increasing temperature the intensity increases at all wavelengths. The maximum of the Planck function, as indicated by the $\times$, shifts to shorter wavelengths with increasing temperature.

## 3.3     Effective temperature of stars

Knowing that the total amount of radiative energy emitted by the ideal light source – the black body – depends on the temperature, we can compare the amount of energy emitted by the stars per cm$^2$ each second with that of a black body. We called the amount of energy emitted by the stars per cm$^2$ s the surface flux $\pi F$. How can we determine the surface flux? We measure the amount of energy arriving above the Earth's atmosphere per cm$^2$. The total amount of energy leaving the star per second must then be given by (if there is no absorption by interstellar matter)

$$L = 4\pi d^2 \times \pi f, \qquad\qquad 3.6$$

where $d$ is the distance between us and the star (see Fig. 3.4). This same amount of energy must leave the star each second, which means

$$L = 4\pi R^2 \times \pi F. \qquad\qquad 3.7$$

Combining equations (3.6) and (3.7), we find

$$\pi F = \pi f \left(\frac{d}{R}\right)^2. \qquad\qquad 3.8$$

$\alpha = R/d$ is the angular radius of the star (see Fig. 3.5), measured in radians. We therefore only have to measure the angular radius of the star in order

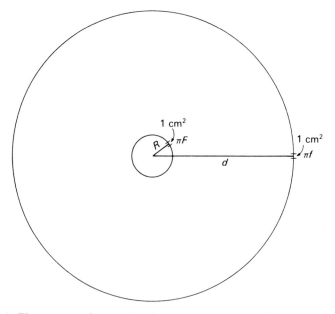

*Fig. 3.4.* The amount of energy leaving the star per second is given by $L = 4\pi d^2 \times \pi f$. It is also equal to $4\pi R^2 \times \pi F$.

to determine $\pi F$. We thus know $\pi F$ if we know $\pi f$, the amount of energy received above the Earth's atmosphere, and the angular radius of the star. For the sun we determined $\pi F_\odot = 6.3 \times 10^{10}$ erg cm$^{-2}$ s$^{-1}$. The relation

$$\pi F = \sigma T_{\text{eff}}^4 \qquad\qquad 3.9$$

defines the *effective temperature* of a star. It gives us the temperature which a black body would need to have if it were to radiate the same amount of energy per cm$^2$ s as the star. For the sun we measure $\pi f = S = 1.37 \times 10^6$ erg cm$^{-2}$ s$^{-1}$. This is called the *solar constant*. With an angular radius of $\alpha_\odot = 959.63$ arcsec, we find $\pi F_\odot = 6.3 \times 10^{10}$ erg cm$^{-2}$ s$^{-1}$, which leads to $T_{\text{eff}_\odot} = 5800$ K for the sun.

For $\alpha$ Lyrae we can also measure the flux $\pi f$ received above the Earth's atmosphere; the angular diameter for $\alpha$ Lyrae was measured by Hanbury-Brown to be $3.2 \times 10^{-3}$ arcsec. From these values and also other measurements we determine that the effective temperature of $\alpha$ Lyrae is 9500 K.

In Table 3.1 we list angular radii and effective temperatures for some representative stars, according to Schmidt-Kaler (1984).

## 3.3    Wien temperatures

The relative energy distribution of a star can also be used to determine a temperature. For instance, we can use Wien's displacement law. For the sun the maximum amount of energy is received at a wavelength of about 5000 Å. From this we also find a temperature of 5800 K for the sun. The agreement of this temperature with the effective temperature gives us hope that these temperatures actually tell us something about the real temperature of the sun in the layer from which we get the radiation.

We can determine so-called color temperatures by comparing, for instance, the $U - B$ colors with that of a black body or the $B - V$ colors. We find that these *color temperatures* are different from the temperatures we have determined so far for the sun (see Fig. 1.7).

In Fig. 1.2 we show the relative energy distribution for $\alpha$ Lyrae as a function of wavelength. The maximum amount of energy is received for a wavelength of about 3800 Å. Using this value, we find a *Wien temperature*

Fig. 3.5. The ratio $R/d$ is the angular radius $\alpha$ of the star if $\alpha$ is measured in radians. $\sin \alpha = \alpha$ because $\alpha \ll 1$.

Table 3.1. *Angular diameters and effective temperatures for some bright stars*

| Star | Spectral type | $(B - V)_0^a$ | $\alpha\ [10^{-3}$ arcsec] $\pm 10\%$ | $T_{\text{eff}}$ |
|---|---|---|---|---|
| *Main sequence stars* | | | | |
| $\xi$ Oph | O9.5 V | −0.30 | 0.51 | 31 900 |
| $\alpha$ Pav | B2.5 V | −0.22 | 0.80 | 17 900 |
| $\alpha$ Leo | B7 V | −0.11 | 1.37 | 12 200 |
| $\alpha$ CMa | A1 V | +0.01 | 5.89 | 10 000 |
| $\alpha$ Lyr | A0 V | +0.00 | 3.24 | 9 500 |
| $\varepsilon$ Sgr | A0 V | −0.03 | 1.44 | 9 500 |
| $\beta$ Leo | A3 V | +0.09 | 1.33 | 8 900 |
| $\alpha$ PsA | A3 V | +0.09 | 2.10 | 8 800 |
| YY Gem | M0.5 V | var | 0.458 | 3 500 |
| *Supergiants* | | | | |
| $\kappa$ Ori | B0.5 Ia | −0.18 | 0.45 | 26 400 |
| $\varepsilon$ Ori | O9.5 Ia | −0.19 | 0.69 | 24 800 |
| $\eta$ CMa | B5 Ia | −0.09 | 0.75 | 13 300 |
| $\beta$ Ori | B8 Ia | −0.03 | 2.55 | 11 500 |
| $\sigma$ CMa | F8 Ia | +0.68 | 3.60 | 6 100 |
| $\alpha$ Sco | M1 Iab | +1.80 | 42.5 | 3 600 |
| 119 Tau | M2 Iab | var | 10.9 | 3 500 |
| $\alpha$ Ori | M2 Iab | +1.86 | 49 | 3 500 |

$^a$ Intrinsic color of the star, i.e., corrected for interstellar reddening (see Chapter 4).

for $\alpha$ Lyrae which is 7600 K. This is much lower than the effective temperature. We suspect that this discrepancy has something to do with the abrupt decrease in energy output seen for $\alpha$ Lyrae for wavelengths shorter than 3700 Å, which looks as if something suddenly absorbs all the energy which was supposed to come out at shorter wavelengths and which might have given us a higher Wien temperature.

We will come back to this question later when we understand better what happens to the radiation in the stellar atmospheres.

## 3.4    Discussion of temperature measurements in stars

We know, of course, that neither the sun nor $\alpha$ Lyrae is a black body. Therefore, we should not be surprised if we find different temperatures from different ways of measuring the temperatures. We will soon see that the temperature of the stars increases inwards. From the comparison with black bodies we nevertheless obtain some estimate for a temperature, which we call the effective temperature $T_{\text{eff}}$. We will have to find out whether this effective temperature has anything to do with the real temperature at the surface of the star. What do we mean by *real* temperature? With real temperature we mean the temperature which describes the motions of the particles correctly. This temperature is often called the kinetic temperature because it described the kinetic properties of the particles. Normally, we do not make this distinction between different temperatures because we assume that we are dealing with just *one* temperature, that all the temperatures which we can find by measuring either the radiation of a body or the motions of the particles are the same. Strictly speaking, this is only true for a body in complete thermodynamic equilibrium, such as a black body. Whenever we have temperature differences we cannot expect different temperature measurements to give the same answer. For instance, if we measure the temperature of the air by a thermometer and at the same time let the sun shine on it, we measure a temperature which is higher than the temperature measured if we prevent the sunlight from reaching the thermometer. The reason is, of course, that the radiation field present in the air but originating in the hot sun belongs to a region with a temperature that is very different from the kinetic temperature of the air. We have no equilibrium between the kinetic temperature of the air molecules and the radiation field. We therefore measure very different temperatures for the given volume of gas, depending on which properties of the gas we use, whether we use the radiation field (by letting the sun shine on the thermometer) or by using the kinetic energy of the molecules, if we exclude the radiation field. In the stars we must expect temperature differences; therefore, we have to be careful with the definition of a temperature.

The example of α Lyrae demonstrates that when we talk about the temperature of a star, we always have to specify how we measure it.

# 4

---

# Basics about radiative transfer

## 4.1 Definition of the intensity $I_\lambda$

At this point, we have to define accurately what we mean by intensity $I_\lambda$. Consider a surface area of size $d\sigma$ and light passing through it perpendicularly in a narrow cone of opening $d\omega$ (see Fig. 4.1).

The amount of energy $E_\lambda$ going per second through the area $d\sigma$ is proportional to $d\sigma$, and to the opening of the cone $d\omega$. The energy $E_\lambda$ passing through this area per second is given by

$$E_\lambda = I_\lambda\, d\omega\, d\sigma\, d\lambda, \qquad \text{or} \qquad I_\lambda = \frac{E_\lambda}{d\omega\, d\sigma\, d\lambda}, \qquad\qquad 4.1$$

i.e., it increases proportionately to the opening of the cone, the size of the area, and the width of the wavelength band $d\lambda$. According to the definition (4.1), the intensity $I_\lambda$ is then the energy which passes per second through an area of $1\,\mathrm{cm}^2$ into a solid angle $\Delta\omega = 1$ in a wavelength band $\Delta\lambda = 1$. Clearly, not all the energy passing into $\Delta\omega = 1$ goes perpendicularly through $d\sigma$, since $\Delta\omega = 1$ describes a rather large cone. Rather, we should think about the energy passing in a narrow cone divided by the size of the solid angle of the cone in the limit of an infinitely small cone opening.

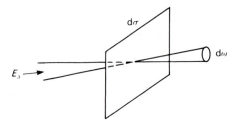

Fig. 4.1. The amount of energy $E_\lambda$ in a wavelength band $\Delta\lambda$ passes through an area $d\sigma$ into a cone with opening $d\omega$.

Let us now consider the energy passing through a surface area $d\sigma$ under an angle $\theta$ with respect to the normal of this surface area (see Fig. 4.2). The effective beam width is reduced by a factor $\cos\theta$, so now

$$E_\lambda = I_\lambda\, d\sigma\, \cos\theta\, d\omega\, d\lambda. \qquad 4.2$$

## 4.2 Radiative energy transport through a gas volume with absorption and emission

We will now derive the radiative transfer equation which describes what happens to a beam of radiation which passes through a volume of gas (see Fig. 4.3). The energy $E_{\lambda_0}$ enters a box with rectangular cross-section.

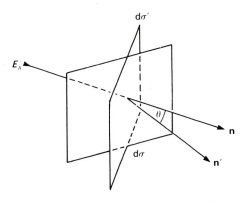

Fig. 4.2. The amount of energy going through $d\sigma$ under an angle $\theta$ is reduced by a factor $\cos\theta$. The amount of energy going through $d\sigma$ is the same as the amount going through the surface area $d\sigma'$, where $d\sigma'$ is the projection of $d\sigma$ onto a plane perpendicular to the beam of light, i.e., $d\sigma' = d\sigma \cos\theta$.

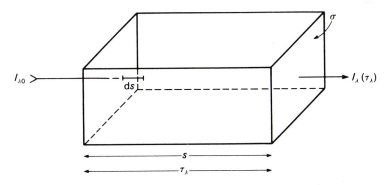

Fig. 4.3. A beam of light passes through a box of cross-section $\sigma$ and length $s$ corresponding to an optical depth $\tau_\lambda = \int_0^s \kappa_\lambda\, ds$. The intensity $I_{\lambda_0}$ entering the box on the left-hand side will be changed by absorption and emission along the path $s$ and will leave the box on the right-hand side with an intensity $I_\lambda$, which depends on the optical depth $\tau_\lambda$ of the box and on its emission properties.

When the light beam passes along the path element $ds$ there is some absorption and the energy $E_\lambda$ is reduced by the amount

$$dE_\lambda = -\kappa_\lambda E_\lambda \, ds = -\kappa_\lambda \, ds \, I_\lambda \, d\omega \, d\lambda \, d\sigma. \qquad 4.3$$

At the same time, there is also some emission from the volume $dV = d\sigma \, ds$, which is given by

$$dE_\lambda = \varepsilon_\lambda \, d\omega \, d\lambda \, ds \, d\sigma, \qquad 4.4$$

where $\varepsilon_\lambda$ now is the amount of energy emitted each second per unit volume into $\Delta\omega = 1$ per wavelength band $\Delta\lambda = 1$. Combining absorption and emission, we find

$$dE_\lambda = dI_\lambda \, d\sigma \, d\lambda \, d\omega = -\kappa_\lambda I_\lambda \, d\omega \, d\lambda \, d\sigma \, ds + \varepsilon_\lambda \, d\omega \, d\lambda \, ds \, d\sigma. \qquad 4.5$$

Dividing by $d\omega \, d\lambda \, d\sigma \, ds$, we obtain in the limit $ds \to 0$

$$\frac{dI_\lambda}{ds} = -\kappa_\lambda I_\lambda + \varepsilon_\lambda. \qquad 4.6$$

This gives

$$\frac{dI_\lambda}{ds} > 0 \text{ if } \varepsilon_\lambda > \kappa_\lambda I_\lambda \qquad \text{and} \qquad \frac{dI_\lambda}{ds} < 0 \text{ if } \varepsilon_\lambda < \kappa_\lambda I_\lambda. \qquad 4.7$$

If we divide equation (4.6) by $\kappa_\lambda$, we obtain

$$\frac{dI_\lambda}{\kappa_\lambda \, ds} = \frac{dI_\lambda}{d\tau_{\lambda s}} = -I_\lambda + \frac{\varepsilon_\lambda}{\kappa_\lambda} = -I_\lambda + S_\lambda, \qquad 4.8$$

where we have introduced the new symbol

$$S_\lambda = \frac{\varepsilon_\lambda}{\kappa_\lambda} \qquad \text{or} \qquad \varepsilon_\lambda = \kappa_\lambda S_\lambda. \qquad 4.9$$

**Equation (4.8) is the radiative transfer equation** in the plane parallel case. $S_\lambda$ is generally called the *source function*. From equation (4.8) we can see that $I_\lambda$ will decrease on the way out if $S_\lambda < I_\lambda$, or $I_\lambda$ will increase if $S_\lambda > I_\lambda$. $I_\lambda$ does not change if $S_\lambda = I_\lambda$.

### 4.3    The source function $S_\lambda$

Let us interpret the physical meaning of the source function $S_\lambda$ by considering the radiative transfer in a black body. According to our earlier definition, this is a volume of gas in perfect thermodynamic equilibrium, in which nothing will change with time. A beam of light passing through such a gas volume therefore will not change either. Under those circumstances, we must then require

$$\frac{dI_\lambda}{ds} = -\kappa_\lambda I_\lambda + \kappa_\lambda S_\lambda = \kappa_\lambda (S_\lambda - I_\lambda) = 0, \qquad 4.10$$

or
$$I_\lambda = S_\lambda. \qquad 4.11$$

Since in a black body $I_\lambda$ equals the Planck function, for which we generally use the symbol $B_\lambda$, we find for a black body or for complete thermodynamic equilibrium

$$S_\lambda = I_\lambda = B_\lambda, \qquad 4.12$$

i.e., **in complete thermodynamic equilibrium, the source function equals the Planck function**, or, in other words, the emissivity

$$\varepsilon_\lambda = \kappa_\lambda B_\lambda, \qquad 4.13$$

which, of course, is the well-known *law of Kirchhoff*. Here

$$B_\lambda = \frac{2hc^2}{\lambda^5} \frac{1}{e^{hc/\lambda kT} - 1}, \qquad 4.14$$

as discussed earlier. Equation (4.7) can now be reformulated as

$$\frac{dI_\lambda}{ds} > 0 \text{ if } S_\lambda > I_\lambda \quad \text{and} \quad \frac{dI_\lambda}{ds} < 0 \text{ if } S_\lambda < I_\lambda. \qquad 4.15$$

If the emission is the same as that of a black body, i.e., if $S_\lambda = B_\lambda$, we can also say

$$\frac{dI_\lambda}{ds} > 0 \text{ if } B_\lambda > I_\lambda \quad \text{and} \quad \frac{dI_\lambda}{ds} < 0 \text{ if } B_\lambda < I_\lambda. \qquad 4.16$$

If in the theory of stellar atmospheres we make the assumption $S_\lambda = B_\lambda$ or $\varepsilon_\lambda = \kappa_\lambda B_\lambda$, this is called the assumption of *local thermodynamic equilibrium*, abbreviated by LTE.

The assumption of LTE does *not* mean we assume complete thermodynamic equilibrium. This would certainly be wrong for the outer layers of the star where we have the large energy loss at the surface. We only assume that the emission of radiation is the same as in a gas in thermodynamic equilibrium at a temperature $T$ corresponding to the temperature of the layer in the stellar atmosphere under consideration. This means, in particular, that even if the absorption were to take place mainly in one spectral line, the emission would still be distributed over all wavelengths according to $B_\lambda$. If LTE holds, the photons always come out at all wavelengths.

## 4.4 Absorption versus emission lines

We shall now discuss under which conditions the spectrum of the light coming out of the box on the right-hand side in Fig. 4.3 will show dark lines on a bright background continuum, and when it will show bright lines on a background of low intensity. In other words, we ask why we see absorption lines in most stellar spectra, while we are accustomed to emission line spectra in the laboratory. In most laboratory experiments, we are dealing with a hot gas in a container, usually glass, or we may even have an open hot gas, such as in the Bunsen burner or free-burning arc. Such a light source will give an emission line spectrum. On the other hand, Kirchhoff and Bunsen found that a container with cold gas in front of a hot light source with a continuous spectrum gives an absorption line spectrum (see Fig. 4.4). For most stars we see absorption lines.

Obviously, most stars correspond to the second case: we see cold gas in front of a hot light source. The deeper layers of the stars are hot, emitting a continuous spectrum, while the outer layers are cool and make the absorption line spectrum. This is a simplified picture, which astrophysicists used about 50 years ago. At that time, they called the cool outer layers the *reversing* layers. We now know that this is a very simplified picture, but the absorption lines clearly indicate outward decreasing temperatures.

We are now in the position to understand quantitatively when absorption lines or emission lines are seen.

Let us talk about the laboratory case first. We base our discussion on equation (4.8):

$$\frac{dI_\lambda}{d\tau_\lambda} = S_\lambda - I_\lambda \qquad \text{with } d\tau_\lambda = \kappa_\lambda \, ds,$$

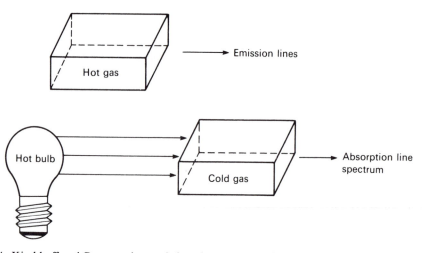

*Fig. 4.4.* Kirchhoff and Bunsen observed that the spectrum of a hot gas shows emission lines, while the spectrum of the light from a hot continuous source, passing through cold gas, shows absorption lines. The wavelengths of the emission and absorption lines are the same if the hot and the cold gas contain the same chemical elements.

giving the change of intensity over a path length $ds$ in Fig. 4.3. We want to integrate this equation in order to find $I_\lambda(\tau_\lambda)$ along the path $s$. In order to achieve this integration, we multiply both sides of the equation by $e^{+\tau_\lambda}$ and obtain

$$\frac{dI_\lambda}{d\tau_\lambda} e^{+\tau_\lambda} + I_\lambda e^{+\tau_\lambda} = S_\lambda e^{+\tau_\lambda}, \tag{4.17}$$

which also can be written as

$$\frac{d}{d\tau_\lambda}(I_\lambda e^{\tau_\lambda}) = S_\lambda e^{\tau_\lambda}. \tag{4.18}$$

This is now a form of equation (4.8), which can easily be integrated over the interval $\tau_\lambda = 0$ to $\tau_\lambda$, giving

$$\int_0^{\tau_\lambda} \frac{d}{d\tau_\lambda}(I_\lambda e^{\tau_\lambda})\, d\tau_\lambda = [I_\lambda e^{+\tau_\lambda}]_0^{\tau_\lambda} = \int_0^{\tau_\lambda} S_\lambda e^{\tau_\lambda}\, d\tau_\lambda = [S_\lambda e^{\tau_\lambda}]_0^{\tau_\lambda}, \tag{4.19}$$

where we have assumed that $S_\lambda$ is constant along the path $s$. With $I_\lambda(\tau_\lambda = 0) = I_{\lambda_0}$ we find that if we insert the values at the boundaries

$$I_\lambda e^{\tau_\lambda} - I_{\lambda_0} = S_\lambda(e^{\tau_\lambda} - 1), \tag{4.20}$$

or, after dividing by $e^{\tau_\lambda}$,

$$I_\lambda = I_{\lambda_0} e^{-\tau_\lambda} + S_\lambda(1 - e^{-\tau_\lambda}). \tag{4.21}$$

The first term on the right-hand side describes the amount of radiation left over from the light entering the box after it has passed through an optical depth $\tau_\lambda$. The second term gives the contribution of the light from the radiation emitted along the path.

Let us first discuss the case that $I_{\lambda_0} = 0$. This corresponds to the upper part of Fig. 4.4. We have only the volume of hot gas; no light is shining on it, but the gas has some emission. In this case, equation (4.21) reduces to

$$I_\lambda = S_\lambda(1 - e^{-\tau_\lambda}) \qquad \text{where in LTE} \qquad S_\lambda = B_\lambda(T). \tag{4.22}$$

Two limiting cases appear to be important:

(a) $\tau_\lambda \ll 1$, i.e., the gas volume has a small optical depth. This is also called the *optically thin case*.

We can then expand the exponential in a Taylor series and obtain

$$e^{-\tau_\lambda} \approx 1 - \tau_\lambda \qquad \text{for } \tau_\lambda \ll 1, \tag{4.23}$$

and find

$$I_\lambda = S_\lambda(1 - 1 + \tau_\lambda) = \tau_\lambda S_\lambda, \tag{4.24}$$

or in LTE, i.e., if $S_\lambda = B_\lambda$,

$$I_\lambda = \tau_\lambda B_\lambda \qquad \text{with } \tau_\lambda = \kappa_\lambda s, \text{ if } \kappa_\lambda \text{ is constant along } s. \tag{4.25}$$

The intensity leaving the box on the other side will be large when $\kappa_\lambda$ is large, and small when $\kappa_\lambda$ is small.

Suppose $\kappa_\lambda$ depends on the wavelength $\lambda$, as shown in Fig. 4.5. The frequencies at which we see the spectral lines are actually frequencies for which the absorption coefficient $\kappa_\lambda$ is very large because they are resonance frequencies in the atom. The wavelengths for very large $\kappa_\lambda$ in Fig. 4.5 therefore correspond to the wavelengths of spectral lines. From equation (4.25) we see that the intensity will be large at those wavelengths where the optical depth is large, i.e., at wavelengths for which $\kappa_\lambda$ is large, which means in the spectral lines. **For $\tau_\lambda \ll 1$ we therefore expect to see emission lines** with large intensity at the wavelengths of large $\kappa_\lambda$.

(b) The other limiting case is the *optically very thick case*, i.e., $\tau_\lambda \gg 1$. In this case, $e^{-\tau_\lambda} \to 0$ and we find from equation (4.22)

$$I_\lambda = S_\lambda, \qquad \text{or in LTE,} \qquad I_\lambda = B_\lambda. \qquad\qquad 4.26$$

In this case, the emitted intensity is independent of $\kappa_\lambda$. We have the situation shown in Fig. 4.6.

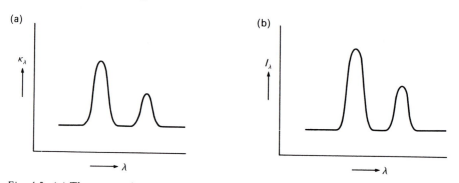

Fig. 4.5. (a) The assumed wavelength dependence of $\kappa_\lambda$. (b) The intensity distribution obtained for $\tau_\lambda \ll 1$ and $I_{\lambda_0} = 0$.

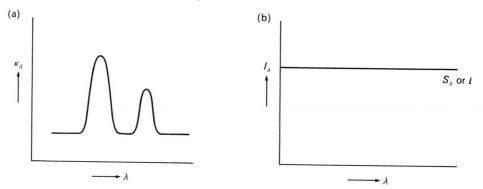

Fig. 4.6. (a) The assumed wavelength dependence of $\kappa_\lambda$. (b) The intensity distribution found for $\tau_\lambda \gg 1$; the intensity is independent of $\kappa_\lambda$.

This is, of course, the case of the black body: $\tau_\lambda \to \infty$ means all the incoming light is absorbed. The emitted intensity is only due to the emission within the box and the intensity observed is given by the source function which, in this case, is the Planck function.

The first case $\tau_\lambda \ll 1$ is important for many astronomical observations. Suppose we see a large interstellar gas cloud or nebula with a density of a few particles per cm$^3$ up to about 10 000 particles per cm$^3$ (see Fig. 4.7). Such a nebula usually shows an emission line spectrum, demonstrating that it is optically thin (see Fig. 4.8), at least in the continuum and in the line wings, not necessarily in the centers of the lines. If in the continuum and in the wings $\tau_\lambda \ll 1$, but in the line centers $\tau_\lambda \gg 1$ we find in the line centers $I_\lambda = S_\lambda$ while in the wings we have $I_\lambda = \tau_\lambda S_\lambda \ll S_\lambda$.

The solar corona also corresponds to the case $\tau_\lambda \ll 1$ and therefore shows an emission line spectrum observable during solar eclipses (see Fig. 4.9). This is due to the small optical depth (and not due to the high temperature, at least not when observed beyond the sun's limb).

Clearly, stars do not show an emission line spectrum because they are optically very thick, $\tau_\lambda \to \infty$. They do not, however, show a smooth intensity distribution for all wavelengths, as does the black body, because the source function $S_\lambda$ is not constant along the path $s$, as we assumed when we derived equation (4.26).

We shall now discuss the case when $I_{\lambda_0} \neq 0$.

Let us again distinguish two cases: the optically thin case with $\tau_\lambda \ll 1$ and the optically thick case with $\tau_\lambda \gg 1$.

In the optically thin case we can again replace $e^{\tau_\lambda}$ by $1 - \tau_\lambda$. Equation (4.21) then reads

$$I_\lambda = I_{\lambda_0}(1 - \tau_\lambda) + S_\lambda \tau_\lambda = I_{\lambda_0} + \tau_\lambda(S_\lambda - I_{\lambda_0}),  \qquad 4.27$$

which for $I_{\lambda_0} = 0$ reduces again to the case discussed previously.

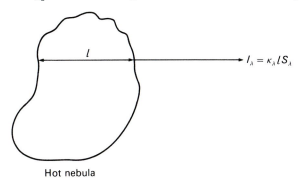

Hot nebula

*Fig. 4.7.* A large interstellar nebula of low density shows an emission line spectrum, indicating that it is optically thin along the line of sight.

For $I_{\lambda_0} \neq 0$ the emerging intensity is given by the original intensity $I_{\lambda_0}$ plus a term whose sign depends on whether $I_{\lambda_0}$ is larger or smaller than $S_\lambda$. If $I_{\lambda_0} > S_\lambda$, then the last term in equation (4.27) is negative. We then rewrite this equation in the form

$$I_\lambda = I_{\lambda_0} - \tau_\lambda(I_{\lambda_0} - S_\lambda).$$
    4.28

This means that from the original intensity there is something subtracted which is proportional to the optical depths $\tau_\lambda$, so there is more intensity

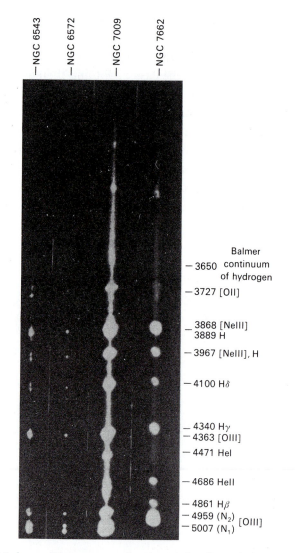

NGC 6543    NGC 6572    NGC 7009    NGC 7662

Balmer
—3650  continuum
      of hydrogen
—3727 [OII]

— 3868 [NeIII]
  3889 H
—3967 [NeIII], H

—4100 Hδ

— 4340 Hγ
—4363 [OIII]
—4471 HeI

—4686 HeII

—4861 Hβ
—4959 (N₂)
—5007 (N₁)  [OIII]

*Fig. 4.8* shows slitless spectra of such emission line nebulae. An image of the nebula is seen for each emission line. This is a positive photo. Bright images are white. (From Unsöld, 1977.)

missing at those wavelengths for which $\kappa_\lambda$ is larger, i.e., in the spectral lines. In this case we will see absorption lines on the background intensity $I_{\lambda_0}$ (see Fig. 4.10). This case is obviously the second case of the Kirchhoff–Bunsen experiment.

In the case that $S_\lambda > I_{\lambda_0}$ we find again for small optical depth

$$I_\lambda = I_{\lambda_0} + \tau_\lambda (S_\lambda - I_{\lambda_0}). \qquad 4.29$$

The last term is now positive. We will see emission lines on top of the background intensity $I_{\lambda_0}$ (see Fig. 4.11).

For $\tau_\lambda \gg 1$ we see from equation (4.21) that in any case $I_\lambda = S_\lambda$, no matter what $I_{\lambda_0}$ was. The original intensity is totally absorbed along the path $s$ with the very large optical depth.

The stars usually show an absorption line spectrum. They must therefore correspond to the case when $I_{\lambda_0}$ – the intensity coming from the deeper layers – is larger than the source function $S_\lambda$ for the top layers of the stars. Let us assume LTE, which means that $S_\lambda = B_\lambda$. For this case the source

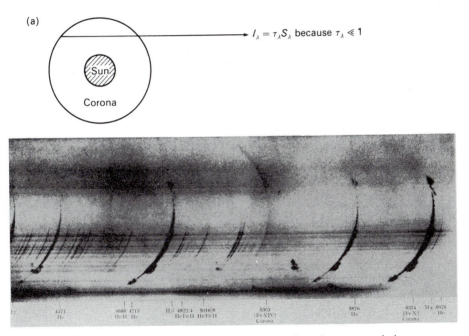

*Fig. 4.9.* (a) The solar corona is optically thin and therefore shows an emission line spectrum. (b) During solar eclipses the optically thin solar corona shows an emission line spectrum. We clearly see the image of the corona in the 5303 Å line emitted by the Fe XIV ion, i.e., an Fe atom which has lost 13 electrons. The other spectral lines seen originate in the solar chromosphere and transition region between chromosphere and corona. The spectrum was obtained by Davidson and Stratton, 1927. This is a negative photo. Bright lines are dark.

function increases with increasing temperature. We can now divide the star into two regions: the optically thick deep layers and the optically thin reversing layer at the top. $I_{\lambda_0}$ entering the reversing layer is then given by the Planck function $B_{\lambda d}$ for the deeper layers, and the source function $S_\lambda$ for the reversing layer is then given by the Planck function $B_{\lambda t}$ for the top layers. The fact that the stars show an absorption line spectrum tells us that the Planck function for the deeper layers is larger than the Planck function for the top layers, i.e.,

$$B_{\lambda d} > B_{\lambda t}, \qquad\qquad 4.30$$

Thus the temperature in the deeper layers is higher than in the top layers.

Our discussion also shows that if we see layers of a star with *temperature increasing outwards, we would expect an emission line spectrum.* This is

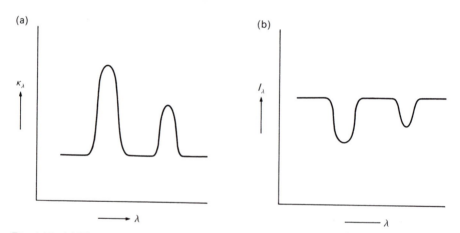

Fig. 4.10. (a) The assumed wavelength dependence of $\kappa_\lambda$. (b) The intensity distribution is shown as a function of wavelength if $\kappa_\lambda(\lambda)$ is given by (a) and $I_{\lambda_0} > S_\lambda$.

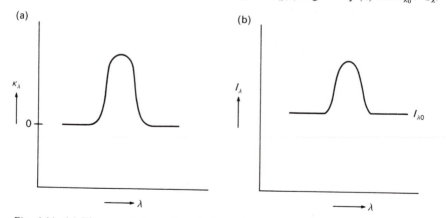

Fig. 4.11. (a) The assumed wavelength dependence of $\kappa_\lambda$. (b) The wavelength dependence of $I_\lambda$ if $\kappa_\lambda(\lambda)$ is given by (a) and $I_{\lambda_0} < S_\lambda$.

actually the case when we observe, for instance, the sun at ultraviolet wavelengths. At the wavelengths longer than about 1700 Å, we still see deeper layers of the photosphere where the temperature is still decreasing outwards, and we see the absorption line spectrum (see Fig. 4.12). For $\lambda < 1600$ Å, the continuous absorption coefficient is so high that we only get light from the chromospheric layers in which the source function is increasing outwards because the temperature increases outwards. For these wavelengths we therefore see an emission line spectrum.

We can now summarize our results.

There are, in principle, two possibilities for the formation of emission line spectra:

1. An optically thin volume of gas in space with no background light, i.e., an emission nebula, emits an emission line spectrum.

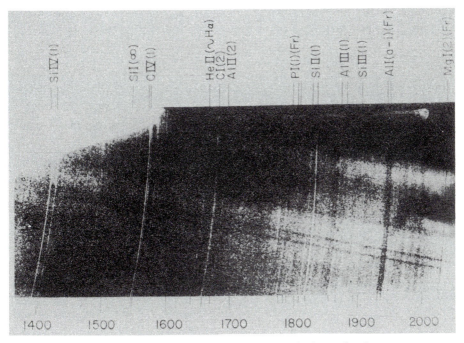

*Fig. 4.12.* The energy distribution for the solar spectrum is shown for the wavelength region 1400 Å $< \lambda <$ 2000 Å. For the longer wavelength region the absorption coefficient in the continuum is still low. We see deeper layers of the photosphere in which the temperature is decreasing outwards. We therefore see an absorption line spectrum. For $\lambda < 1600$ Å the continuous absorption coefficient is much larger. For these wavelengths we see chromospheric layers in which the temperature is increasing outwards. We see an emission line spectrum. (From Zirin, 1966, p. 157.) This is a positive photo. Bright lines are white.

2. An optically thick volume of gas in which the source function is increasing outwards. The solar spectrum in the ultraviolet is an example.

An absorption line spectrum is formed in an optically thick gas in which the source function decreases outwards, which generally means that the temperature is decreasing outwards. (If the source function is not equal to the Planck function, then it can also happen that the source function decreases outwards while the temperature increases outwards. Such exceptions have to be kept in mind.)

An absorption line spectrum can also form if an optically thin gas is penetrated by a background radiation whose intensity $I_{\lambda_0}$ is larger than the source function. A thin shell around a star might be an example, or the interstellar medium between us and the star.

# 5

## Radiative transfer in stellar atmospheres

### 5.1    The radiative transfer equation

In the outer layers of the stars, heat transport must be by radiation, since there is no other means of transporting heat into the vacuum surrounding the star. (The extremely low density interstellar material cannot provide any other method of heat transport comparable to the radiative energy loss of a star. It therefore can be considered to be a vacuum in this context.) These outer layers, where heat transport is by radiation only, may be shallow in some stars and may extend almost to the center of the star in others, as we shall see later. In the following discussion, we shall consider the case in which these outer layers have a very large optical depth $\tau_\lambda$ (see Chapter 1) but a geometrical height which is small compared to the radius of the star. In the case of the sun, for instance, a layer with $\tau_\lambda = 10$ at visual wavelength has a geometrical height of about 500 km, which is certainly small in comparison with the solar radius of 700 000 km. In this case the radius of curvature is much larger than the height of the layer and we can consider it as plane parallel, as we did with the Earth's atmosphere in Chapter 1.

As in Chapter 1, we consider the flow of radiative energy through this outer layer, which we call the atmosphere of the star. In contrast to the Earth's atmosphere, there is no solid star underneath, since the stellar temperatures are much too high to permit solidification – except perhaps in such exotic objects as the white dwarfs.

Fig. 5.1 explains the geometry of the radiation beam passing through a plane parallel atmosphere under an angle $\vartheta$. Emission occurs in all the volume elements $d\sigma\,ds$ along the path $s$. Radiation with intensity $I_\lambda$ at wavelength $\lambda$ passes through the atmosphere under an angle $\vartheta$ with respect to the normal $\mathbf{n}$ of the atmosphere. In principle, the intensity could depend also on the angle $\varphi$. However, in the case of our nice, spherically symmetric stars, there is no reason why it should.

Along its path $s$ the intensity will change by absorption and emission:

$$\frac{dI_\lambda(\vartheta)}{ds} = -\kappa_\lambda I_\lambda(\vartheta) + \varepsilon_\lambda = -\kappa_\lambda I_\lambda(\vartheta) + \kappa_\lambda S_\lambda. \qquad 5.1$$

With $\kappa_\lambda \, ds = d\tau_{\lambda s}$ we can introduce the optical depth $\tau_{\lambda s}$ along $s$ for the outgoing beam of light, which will depend then on the direction $\vartheta$ of the beam. As in Chapter 1, we introduce the optical depth $\tau_\lambda$ perpendicular to the stellar surface going from the outside in. From Fig. 5.2 we see that

$$\cos\vartheta = -\frac{d\tau_\lambda}{d\tau_{\lambda s}} \qquad \text{or} \qquad d\tau_{\lambda s} = -d\tau_\lambda \sec\vartheta. \qquad 5.2$$

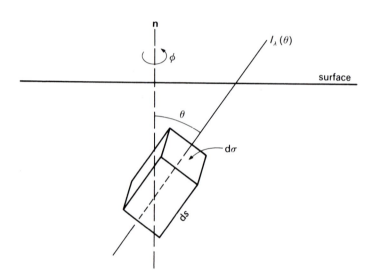

Fig. 5.1. A light beam passing through the stellar atmosphere from the inside to the outside under an angle $\vartheta$ with respect to the normal **n** to the surface experiences absorption along the path $s$ and also emission.

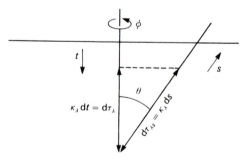

Fig. 5.2 shows the relation between the optical depth $\tau_{\lambda s}$ along the outgoing light beam, leaving the surface under an angle $\vartheta$, and the optical depth $\tau_\lambda$ going from the surface in, perpendicular to the surface.

Introducing this into the transfer equation (5.1) gives

$$\cos\vartheta\frac{dI_\lambda(\vartheta)}{d\tau_\lambda} = I_\lambda(\vartheta) - S_\lambda, \qquad \text{where } I_\lambda = I_\lambda(\tau_\lambda, \vartheta), \qquad 5.3$$

where all signs have changed, since we are now looking from the outside in, i.e. opposite to the direction of light propagation.

Equation (5.3) is the *radiative transfer equation* for the plane parallel case. Whenever we talk about changes in intensity due to absorption or emission, this is the equation which describes what is happening. This transfer equation enables us to derive an integral relation between the intensity leaving the stellar surface and the depth dependence of the source function. In order to determine $I_\lambda(\tau_\lambda, \vartheta)$ we need to integrate equation (5.3).

## 5.2    The surface intensities

In order to derive the intensity $I_\lambda(0, \vartheta)$ at the surface, we multiply equation (5.3) by $e^{-\tau_\lambda \sec\vartheta}$, similar to the derivation of equation (4.17), and obtain

$$\frac{dI_\lambda(\vartheta)}{\sec\vartheta\, d\tau_\lambda} e^{-\tau_\lambda \sec\vartheta} = I_\lambda(\vartheta) e^{-\tau_\lambda \sec\vartheta} - S_\lambda e^{-\tau_\lambda \sec\vartheta} \qquad 5.4$$

where $S_\lambda = S_\lambda(\tau_\lambda)$ and $I_\lambda(\vartheta) = I_\lambda(\tau_\lambda, \vartheta)$. This can be written as

$$\frac{d(I_\lambda e^{-\tau_\lambda \sec\vartheta})}{d(\tau_\lambda \sec\vartheta)} = -S_\lambda e^{-\tau_\lambda \sec\vartheta}. \qquad 5.5$$

In order to derive the intensity $I_\lambda(0, \vartheta)$ at the surface, we now integrate $d(\tau_\lambda \sec\vartheta)$ from 0 to $\infty$, and find

$$[I_\lambda(\vartheta)e^{-\tau_\lambda \sec\vartheta}]_0^\infty = -\int_0^\infty S_\lambda e^{-\tau_\lambda \sec\vartheta}\, d(\tau_\lambda \sec\vartheta) \qquad 5.6$$

and

$$I_\lambda(0, \vartheta) = \int_0^\infty S_\lambda(\tau_\lambda)e^{-\tau_\lambda \sec\vartheta}\, d(\tau_\lambda \sec\vartheta). \qquad 5.7$$

where we have made use of the fact that, for $\tau_\lambda \to \infty$, $e^{-\tau_\lambda \sec\vartheta} \to 0$.

Equation (5.7) can easily be interpreted if we remember that the emission along $ds^*$ is determined by $\varepsilon_\lambda\, dV = \varepsilon_\lambda\, ds = S_\lambda \kappa_\lambda\, ds = S_\lambda \kappa_\lambda\, d(t \sec\vartheta) = S_\lambda\, d(\tau_\lambda \sec\vartheta)$, where $d(\tau_\lambda \sec\vartheta) = d\tau_{\lambda s}$, i.e., the optical depth along the path $s$. From this emission at $\tau_\lambda$ the fraction $S_\lambda\, d(\tau_\lambda \sec\vartheta)e^{-\tau_\lambda \sec\vartheta}$ arrives at the surface; the rest has been absorbed on the way. $I_\lambda(0, \vartheta)$ results from the summation of all the contributions of the volume elements along the path of the light as expressed by equation (5.7).

Let us evaluate $I_\lambda(0, \vartheta)$ for a simple case of $S_\lambda(\tau_\lambda)$. Assume

$$S_\lambda(\tau_\lambda) = a_\lambda + b_\lambda \tau_\lambda. \qquad 5.8$$

---

* Remember $I_\lambda$ is the energy going through $1\,cm^2$, therefore $d\sigma = 1\,cm^2$ and $dV = 1\,cm^2\, ds = ds$.

We derive

$$I_\lambda(0, \vartheta) = \int_0^\infty a_\lambda e^{-\tau_\lambda \sec \vartheta}\, d(\tau_\lambda \sec \vartheta) + b_\lambda \int_0^\infty \tau_\lambda e^{-\tau_\lambda \sec \vartheta}\, d(\tau_\lambda \sec \vartheta)$$

$$= A_i + B_i \cos \vartheta = a_\lambda + b_\lambda \cos \vartheta = S_\lambda(\tau_\lambda = \cos \vartheta). \qquad 5.9$$

In the general case, when

$$S_\lambda(\tau_\lambda) = \sum_i a_{\lambda i} \tau_\lambda^i, \qquad 5.10$$

we find

$$I_\lambda(0, \vartheta) = \sum_i A_i \cos^i \vartheta, \qquad \text{with } A_i = a_{\lambda i} i! \qquad 5.11$$

From (5.9), we derive

$$I_\lambda(0, \vartheta) = a_\lambda + b_\lambda \cos \vartheta = S_\lambda(\tau_\lambda = \cos \vartheta), \qquad 5.12$$

where $\tau_\lambda$ is the optical depth perpendicular to the surface. The angular distribution of $I_\lambda(0, \vartheta)$ for different values of $a_\lambda$ and $b_\lambda$ is shown in Fig. 5.3. For $\vartheta = 0$, $I_\lambda = a_\lambda + b_\lambda$, while for $\vartheta = 90°$, $I_\lambda = a_\lambda$. How can we observe this angular distribution? If we as observers are at a very large distance from the star, we receive essentially parallel light beams. Fig. 5.4 shows then that for the center of the star we see radiation leaving the star

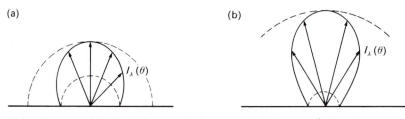

Fig. 5.3. Polar diagram of $I_\lambda(\vartheta)$, (a) for $a_\lambda = 1$, $b_\lambda = 1$, and (b) for $a_\lambda = \frac{1}{2}$, $b_\lambda = 2$. The larger $b_\lambda/a_\lambda$, the more the radiation is concentrated in the forward direction.

Fig. 5.4. Light beams received from a star by a distant observer are essentially parallel. The light coming from the center of the stellar disc leaves the surface perpendicularly to the surface ($\vartheta = 0$), while the light observed at the limb leaves the surface essentially parallel to the surface ($\vartheta = 90°$). The angular distribution of $I_\lambda(\vartheta)$ can be observed for the sun by means of the center-to-limb variation of the intensity of the solar disc.

perpendicular to the surface, while at the limb the starlight which we receive leaves the surface under an angle $\vartheta = 90°$. Therefore, in the center we should see an intensity $I_\lambda = a_\lambda + b_\lambda$, while at the limb we should see only $I_\lambda = a_\lambda$. We get less light from the limb than from the center.

Unfortunately, the sun is the only star for which we can see the center-to-limb variation of the intensity. In Fig. 5.5 we show a photograph of the solar disc. The limb darkening is obvious.

Equation (5.12) tells us that in the center of the solar disc ($\vartheta = 0$, $cos\ \vartheta = 1$) we see the source function $S_\lambda$ at a depth $\tau_\lambda = 1$, at the edge of the disc ($\vartheta = \frac{1}{2}\pi$, $\cos\ \vartheta = 0$) we see $S_\lambda$ at the surface. Actually, since $t/s = \cos\ \vartheta$ in the plane parallel approximation we see that $\tau_\lambda = \cos\ \vartheta$ if $\tau_{\lambda s} = 1$. Equation (5.12) then means that we always see into the atmosphere down to a depth where $\tau_{\lambda s} = 1$, which corresponds to $\tau_\lambda = \cos\ \vartheta$ (see Fig. 5.6).

## 5.3    The fluxes

Usually, what we need to know is the total amount of energy leaving $1\ \mathrm{cm}^2$ per $\Delta\lambda = 1$ per second, which we called $\pi F_\lambda$ (see Fig. 5.9b). The

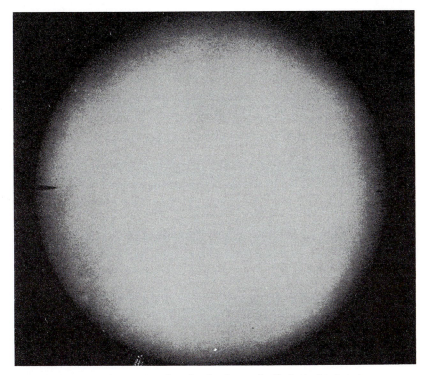

*Fig. 5.5.* This photograph of the solar disc clearly shows the limb darkening due to the angular dependence of the intensity $I_\lambda(\vartheta)$. (From Abell, 1982, p. 512.)

amount of energy going through 1 cm$^2$ per second per $\Delta\lambda = 1$ into the solid angle $d\omega$ in the direction inclined by the angle $\vartheta$ to the normal of the area 1 cm$^2$ is given by (see Fig. 4.2)

$$dE_\lambda = I_\lambda \cos\vartheta\, d\omega,$$   5.13

and

$$\pi F_\lambda = \int I_\lambda \cos\vartheta\, d\omega,$$   5.14

integrated generally over the whole solid angle, $\omega = 4\pi$.

For this integration we must express $d\omega$ by means of $\vartheta$ and $\varphi$, where $\varphi$ corresponds to longitude. How do we describe the opening of the cone $d\omega$? It is best described by the area which it cuts out of the surface of a sphere with radius 1. From Fig. 5.7 we see that the surface area $d\sigma$ at the

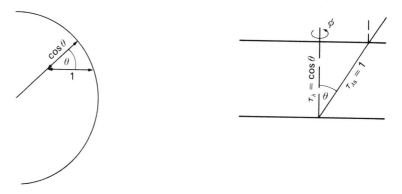

Fig. 5.6. The radial optical depth equals cos $\vartheta$ if the optical depth along the line of sight is 1.

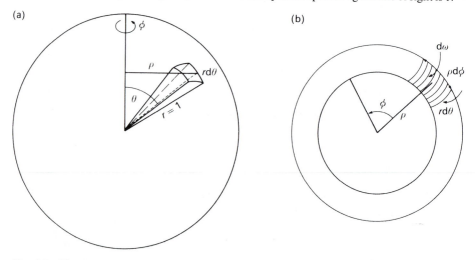

Fig. 5.7. The description of $d\omega$ in polar coordinates. (a) Sphere viewed from the side, (b) sphere viewed from the top.

distance (colatitude) $\vartheta$ from the pole, corresponding to a change $d\varphi$ in longitude and $d\vartheta$ in colatitude, is given by

$$d\sigma = d\omega = r\rho \, d\vartheta \, d\varphi; \qquad \text{with } \rho = r \sin \vartheta = \sin \vartheta. \qquad 5.15$$

For $r = 1$ we find

$$d\sigma = d\omega = r^2 \sin \vartheta \, d\vartheta \, d\varphi = \sin \vartheta \, d\vartheta \, d\varphi. \qquad 5.16$$

With this we find

$$\pi F_\lambda(\tau_\lambda) = \int_0^{2\pi} \int_0^\pi I_\lambda(\tau_\lambda, \vartheta) \cos \vartheta \sin \vartheta \, d\vartheta \, d\varphi. \qquad 5.17$$

For $\varphi$ going from 0 to $2\pi$, we cover the whole sphere if $\vartheta$ goes from 0 to $\pi$. Remembering that $d(\cos \vartheta)/d\vartheta = -\sin \vartheta$ or $d(\cos \vartheta) = -\sin \vartheta \, d\vartheta$, we can rewrite the above equations as

$$\pi F_\lambda(\tau_\lambda) = -\int_0^{2\pi} \int_1^{-1} I_\lambda(\tau_\lambda, \vartheta) \cos \vartheta \, d(\cos \vartheta) \, d\varphi$$

$$= \int_0^{2\pi} \int_{-1}^1 I_\lambda(\tau_\lambda, \vartheta) \cos \vartheta \, d(\cos \vartheta) \, d\varphi. \qquad 5.18$$

In the plane parallel case, we do not find any dependence of $I_\lambda$ on the longitude $\varphi$. We can therefore easily integrate over $\varphi$, which yields a factor $2\pi$, and obtain

$$\pi F_\lambda(\tau_\lambda) = 2\pi \int_{-1}^1 I_\lambda(\tau_\lambda, \vartheta) \cos \vartheta \, d(\cos \vartheta). \qquad 5.19$$

## 5.4    The surface flux $\pi F(0)$ and the effective temperature

Let us now derive $\pi F_\lambda$ on the surface, i.e., for $\tau_\lambda = 0$. It is best to split the integral in equation (5.19) into two parts

$$\vartheta > \tfrac{1}{2}\pi \qquad\qquad\qquad \vartheta < \tfrac{1}{2}\pi$$

i.e., radiation going inward     i.e., radiation going outward

At the surface, no radiation will come from the outside. This means $I_\lambda(0, \vartheta) = 0$ for $\tfrac{1}{2}\pi < \vartheta < \pi$. Therefore, the first integral is 0, and we find

$$\pi F_\lambda(0) = 2\pi \int_0^1 I_\lambda(0, \vartheta) \cos \vartheta \, d(\cos \vartheta). \qquad 5.20$$

As in Section 5.2, we now approximate the depth dependence of the source function by equation (5.8):

$$S_\lambda(\tau_\lambda) = a_\lambda + b_\lambda \tau_\lambda.$$

We then find equation (5.9):

$$I_\lambda(0, \vartheta) = a_\lambda + b_\lambda \cos \vartheta.$$

Inserting this into equation (5.20), we derive

$$\pi F_\lambda(0) = 2\pi \int_0^1 (a_\lambda + b_\lambda \cos \vartheta) \cos \vartheta \, d(\cos \vartheta) = (a_\lambda + \tfrac{2}{3}b_\lambda)\pi. \qquad 5.21$$

Comparing this with our source function approximation (5.8), we see that

$$F_\lambda(0) = S_\lambda(\tau_\lambda = \tfrac{2}{3}). \qquad 5.22$$

This is the important *Eddington–Barbier relation*. It is very helpful for the understanding of the formation of stellar spectra. The flux which comes out of the stellar surface equals the source function at the optical depth $\tau_\lambda = \tfrac{2}{3}$. This is accurate to the degree that (5.8) is a good approximation.

We now again assume LTE, i.e., that $S_\lambda(\tau_\lambda) = B_\lambda(\tau_\lambda)$. We then find

$$F_\lambda(0) = B_\lambda(\tau_\lambda = \tfrac{2}{3}). \qquad 5.23$$

For simplicity, let us now further assume that the absorption coefficient $\kappa_\lambda$ is *independent of* $\lambda$, i.e., that it has the same value for all wavelengths $\lambda$. We call such a (hypothetical) atmosphere a *grey atmosphere*. For a grey atmosphere we find with $\kappa_\lambda = \kappa$ and $\tau_\lambda = \tau$ that

$$F_\lambda(0) = B_\lambda(T(\tau = \tfrac{2}{3})). \qquad 5.24$$

This means the energy distribution of $F_\lambda$ is that of a black body corresponding to the temperature at the optical depth $\tau = \tfrac{2}{3}$. Then

$$\pi F(0) = \pi \int_0^\infty F_\lambda(0) \, d\lambda = \pi \int_0^\infty B_\lambda(T_0) \, d\lambda, \qquad 5.25$$

or

$$\pi F(0) = \sigma T_0^4 \qquad \text{with } T_0 = T(\tau = \tfrac{2}{3}). \qquad 5.26$$

Since, by definition,

$$\pi F(0) = \sigma T_{\text{eff}}^4 \qquad \text{we find} \qquad T_{\text{eff}}^4 = T^4(\tau = \tfrac{2}{3}), \qquad 5.27$$

or **the temperature at $\tau = \tfrac{2}{3}$ must equal the effective temperature**. This result must hold if our assumptions hold, that is, if we have a linear depth dependence of the source function and if we have a grey atmosphere. Both assumptions are not strictly true, but they turn out to be quite good approximations, at least in many cases, as, for instance, in the solar case where we saw that the energy distribution represents that of a black body with $T = T_{\text{eff}}$ rather well.

## 5.5     The flux $F$ and the anisotropy of the radiation field

Let us go back to equation (5.19) for a moment. Suppose $I_\lambda$ does not depend on $\vartheta$. Then it can, as a constant, be taken in front of the integral and we are left with

$$\pi F_\lambda = 2\pi I_\lambda \int_{-1}^{+1} \cos \vartheta \; \mathrm{d}(\cos \vartheta) = 0, \qquad\qquad 5.28$$

i.e., in an isotropic radiation field the radiative flux is zero (Fig. 5.8). In the integral of (5.18) we have $\cos \vartheta$ as a factor which changes sign for $\vartheta > \frac{1}{2}\pi$. This means the contribution of the radiation from $\vartheta > \frac{1}{2}\pi$ is subtracted from the contribution for $\vartheta < \frac{1}{2}\pi$.

The flux integral (5.19) measures the anisotropy of the radiation field. Of course, the flux is in principle a vector with a direction, depending on the $\vartheta = 0$ direction, as can be seen from Fig. 5.9(a) and (b).

In the spherically symmetric case of the stars, however, we will generally expect the flux to go radially outward, and therefore it has only a radial component. So we will only talk about the radial component of the flux and call that $F$ or $F_\lambda$. In the spherically symmetric star the other components are zero.

The flux obviously describes the net amount of energy flowing outwards.

Let us now ask whether the observations of stellar spectra give us the intensity distribution $I_\lambda(0, \vartheta)$ or the flux $F_\lambda(0)$ as a function of $\lambda$.

Since we cannot see the different parts of the star separately (except for the sun), we can only observe the total amount of light coming from all parts of the visible hemisphere of a star.

The amount of radiation leaving $1 \, \mathrm{cm}^2$ each second is given by $I_\lambda \cos \vartheta \, \Delta\omega$. For a given angle $\vartheta$, we get radiation from an area $2\pi\rho R \, \mathrm{d}\vartheta$ with $\rho = R \sin \vartheta$ (compare Fig. 5.10).

*Fig. 5.8.* In an isotropic radiation field, the net radiative flux is zero.

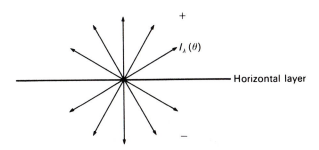

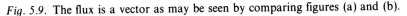

*Fig. 5.9.* The flux is a vector as may be seen by comparing figures (a) and (b).

The solid angle $\Delta\omega$, out of which we receive the measured amount of energy per cm$^2$ at the Earth, is determined by the distance of the Earth and is given by $\Delta\omega = 1\,\text{cm}^2/d^2$ (see Fig. 5.11). The total amount of energy received by $1\,\text{cm}^2$ per second at the Earth is then

$$E_\lambda = \Delta\omega \int_0^{\pi/2} I_\lambda(0, \vartheta) \cos\vartheta \times 2\pi R^2 \sin\vartheta \, d\vartheta$$

$$= 2\pi R^2 \, \Delta\omega \int_0^{\pi/2} I_\lambda(0, \vartheta) \cos\vartheta \sin\vartheta \, d\vartheta. \qquad 5.29$$

Comparing (5.29) with equation (5.20) we see that

$$E_\lambda = 2\pi R^2 \times \tfrac{1}{2} F_\lambda(0) \Delta\omega = \pi R^2 F_\lambda(0) \Delta\omega, \qquad 5.30$$

or, considering the star to be a luminous disc of area $\pi R^2$, we can compute

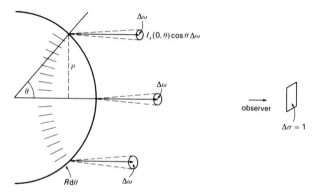

Fig. 5.10. Stellar observations give us the sum of the intensities leaving the stellar surface under all angles of $\vartheta$ between 0 and $\tfrac{1}{2}\pi$. The dashed region indicates the area with $\vartheta$ between $\vartheta$ and $\vartheta + d\vartheta$.

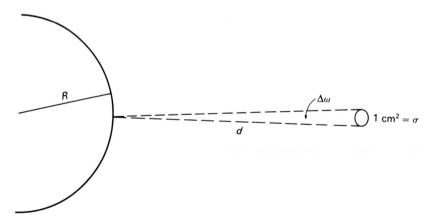

Fig. 5.11. The solid angle $\Delta\omega$ out of which we receive radiation from the star with a receiver of area $\sigma$ is given by $\sigma/d^2$. This is the area that $\Delta\omega$ would cut out of a sphere with radius 1. For $d \gg R$, $\Delta\omega$ is the same for all points on the stellar surface.

the mean intensity coming from $1\,\text{cm}^2$ of this disc as

$$\bar{I}_\lambda = \frac{E_\lambda}{\Delta\omega\pi R^2} = 2\int_0^{\pi/2} I_\lambda(0,\vartheta)\cos\vartheta\sin\vartheta\,d\vartheta = F_\lambda(0).\qquad 5.31$$

$F_\lambda(0)$ is the mean intensity radiated from the stellar disc.

*The spectra of stars give us the energy distribution of the flux $F_\lambda$.* While the flux by definition describes the energy coming from $1\,\text{cm}^2$ and going into all angles $\vartheta$ and $\varphi$, we see from the star's hemisphere the radiation for different $\vartheta$ and $\varphi$ coming from different areas of the stellar surface. The net effect is the same as if we could see all directions from the same spot.

## 5.6    The radiation density

The mean intensity, $J_\lambda$, of a radiation field is defined by

$$J_\lambda = \int I_\lambda \frac{d\omega}{4\pi},\qquad 5.32$$

which for an isotropic radiation field gives $I_\lambda = J_\lambda$, as it must be.

If the intensity $I_\lambda(\vartheta,\varphi)$ is known, we might ask how large is the amount of radiative energy present at any given moment in a volume element of $1\,\text{cm}^3$? We call this the radiation density $u_\lambda$ or $u_\nu$, depending on whether we refer to a wavelength band of $1\,\text{cm}$ or a frequency band of 1 hertz. In order to find the radiation density $u_\lambda$ we have to ask how much energy enters a surface element of the volume under consideration per second. We then have to integrate over the whole surface and consider that within one second each beam is spread out over a distance corresponding to the velocity of light $c$. We follow Unsöld's (1955) derivation.

In Fig. 5.12 we consider a small volume $\Delta V$. Through the surface element $d\sigma$ an amount of energy $dE$ enters this volume per second, at an angle $\vartheta$

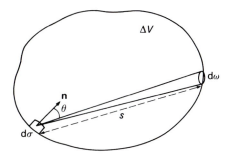

*Fig. 5.12.* The intensities $I_\lambda(\vartheta,\varphi)$ enter the volume element $\Delta V$ through a surface element $d\sigma$. At any given moment the fraction $I_\lambda\,d\sigma\cos\vartheta\,d\omega \times (s/c)$ is to be found in the volume of gas under consideration.

to the normal **n** on d$\sigma$. The amount is

$$dE = I_\lambda(\vartheta, \varphi) \cos \vartheta \, d\sigma \, d\omega. \qquad 5.33$$

In each second this energy is spread out over a distance $c$. At any moment a fraction $s/c$ is actually in the volume $\Delta V$, where $s$ is the path length of the beam through $\Delta V$. At any given time the amount of energy in $\Delta V$ is therefore given by

$$\Delta V u_\lambda = \int_\omega \int_\sigma I_\lambda(\vartheta, \varphi) \, d\sigma \cos \vartheta \, d\omega \, \frac{s}{c}. \qquad 5.34$$

With d$\sigma \cos \vartheta \, s = dV$ we find after integration over $dV$ for the radiation density

$$u_\lambda = \frac{1}{c} \int_\omega I_\lambda(\vartheta, \varphi) \, d\omega, \qquad 5.35$$

or

$$u_\lambda = \frac{4\pi}{c} J_\lambda. \qquad 5.36$$

# 6

# The depth dependence of the source function

## 6.1 Empirical determination of the depth dependence of the source function for the sun

As we have seen in Chapter 5, the intensity leaving the stellar surface under an angle $\vartheta$ is given by

$$I_\lambda(0, \vartheta) = \sum_i A_{\lambda i} \cos^i \vartheta \qquad \text{if } S_\lambda(\tau_\lambda) = \sum_i a_{\lambda i} \tau_\lambda^i, \qquad 6.1$$

where the relation between $A_{\lambda i}$ and $a_{\lambda i}$ is known to be (see (5.11))

$$A_{\lambda i} = a_{\lambda i} i! \qquad 6.2$$

If we can measure $A_{\lambda i}$ for a given wavelength $\lambda$, i.e., if we measure the center-to-limb variation of $I_\lambda(0, \vartheta)$, then we can determine the depth dependence of the source function. We can do this for all wavelengths $\lambda$ and obtain $S_\lambda(\tau_\lambda)$ for all $\lambda$.

As an example, we have chosen the wavelength $\lambda = 5010$ Å. Measured data for the center-to-limb variation can be approximated by (see the table in Chapter 6, problem 1, page 233).

$$I_\lambda(0, \vartheta) = [a_0(\lambda) + a_1(\lambda) \cos \vartheta + 2a_2(\lambda) \cos^2 \vartheta] I_\lambda(0, 0). \qquad 6.3$$

From this expression we find, according to (6.1), that

$$S_\lambda(\tau_\lambda(5010 \text{ Å})) = [a_0(\lambda) + a_1(\lambda)\tau_\lambda(5010 \text{ Å}) + a_2(\lambda)\tau_\lambda^2(5010 \text{ Å})] I_\lambda(0, 0). \quad 6.4$$

In Fig. 6.1 we have plotted the source function $S_\lambda(\tau_\lambda(5010 \text{ Å}))$ as a function of the optical depth according to equation (6.4). If we now go one step further and assume again that $S_\lambda = B_\lambda(T(\tau_\lambda))$, then we know at each optical depth $\tau_\lambda$ how large the Planck function $B_\lambda$ has to be. Since $B_\lambda$ depends only on the temperature $T$ and on the wavelength $\lambda$ we have one equation at each optical depth to determine the temperature $T$, which will give the required value for the Planck function, i.e.,

$$S_\lambda(\tau_\lambda) = B_\lambda(T(\tau_\lambda)) = \frac{2hc^2}{\lambda^5} \frac{1}{e^{hc/\lambda kT} - 1}, \qquad 6.5$$

where, in our example, the wavelength $\lambda$ is 5010 Å.

From equation (6.5) and the values of $S_\lambda(\tau_\lambda)$ derived from the center-to-limb variation, the temperature $T$ can be determined as a function of the optical depth for the wavelength of 5010 Å.

The optical depth dependence of the temperature obtained in this way for the wavelength of 5010 Å is shown in Fig. 6.2. As we expected, the temperature increases with increasing optical depths.

## 6.2    The wavelength dependence of the absorption coefficient in the sun

There is, of course, nothing magic about the wavelength of 5010 Å. We can follow the same procedure for any other wavelength $\lambda$. For each wavelength we find a relation $T(\tau_\lambda)$, showing the dependence of the temperature on that particular optical depth for the chosen wavelength. In Fig. 6.3 we show the results for a number of wavelengths. We find different curves for the different wavelengths because $\kappa_\lambda$ is dependent on $\lambda$. For a given layer in the solar atmosphere with a given geometrical depth

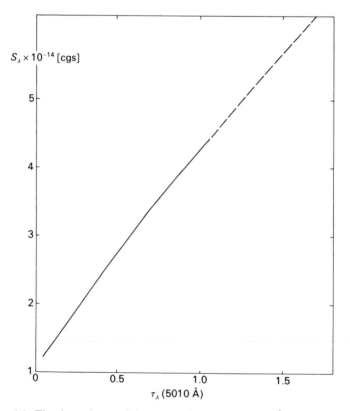

*Fig. 6.1.* The dependence of the source function $S_\lambda$ (5010 Å) on the optical depth $\tau_\lambda$ for the wavelength of 5010 Å. From $\tau_\lambda > 1$ we receive little radiation. The values for deeper layers therefore become less certain as indicated by the dashed line.

$t$ and a particular temperature, we have different optical depths, $\tau_\lambda$, for different wavelengths $\lambda$. We know, however, that at a given geometrical depth $t$ we must have *one* value for the temperature. All points with the same temperature $T_0$ must therefore refer to the same geometrical depth. We draw a horizontal line through Fig. 6.3 which connects all the points with the temperature $T_0 = 6300$ K. All these points must belong to the same geometrical depth $t$. We now write the optical depths $\tau_\lambda$ in the form

$$\tau_\lambda = \kappa_\lambda t. \qquad 6.6$$

Here $\kappa_\lambda$ is an average over depth down to depth $t$. For all the points on the horizontal line we know that the value of $t$ is the same. We read off at the abscissa the optical depths $\tau_{\lambda_1}, \tau_{\lambda_2}, \tau_{\lambda_3}$, etc., which belong to this depth $t$. Equation (6.6) now tells us that

$$t(6300 \text{ K}) = \frac{\tau_{\lambda_3}}{\kappa_{\lambda_3}} = \frac{\tau_{\lambda_2}}{\kappa_{\lambda_2}} = \frac{\tau_{\lambda_1}}{\kappa_{\lambda_1}}, \qquad 6.7$$

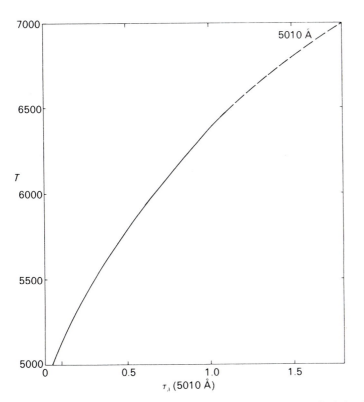

*Fig. 6.2.* The dependence of the solar temperature $T$ on the optical depth at $\lambda = 5010$ Å is shown as derived from the observed center-to-limb variation of the solar intensity at $\lambda = 5010$ Å. For $\tau_\lambda > 1$ the values become less certain, as indicated by the dashed line.

or

$$\frac{\tau_{\lambda_3}}{\tau_{\lambda_2}} = \frac{\kappa_{\lambda_3}}{\kappa_{\lambda_2}} \quad \text{and} \quad \frac{\tau_{\lambda_3}}{\tau_{\lambda_1}} = \frac{\kappa_{\lambda_3}}{\kappa_{\lambda_2}}, \qquad\qquad 6.8$$

which means we can determine the ratio of all $\kappa_{\lambda_i}$ with respect to one $\kappa_{\lambda_0}$.

For a given temperature $T(t)$ we can also plot the values $\tau_\lambda(T)$ as a function of $\lambda$. Because $t$ is the same for all $\lambda$, the wavelength dependence of $\tau_\lambda$ shows directly the wavelength dependence of $\kappa_\lambda$, as was first derived by Chalonge and Kourganoff in 1946.

If $\kappa_{\lambda_i}$ does depend on the geometrical depth $t$, we will find an average value of $\kappa_{\lambda_i}$ down to the depth $t$. For different temperatures, i.e., for different depths, these averages will be somewhat different. The results are shown in Fig. 6.4.

It turns out that the wavelength dependence of $\kappa_\lambda$ agrees with the absorption coefficient which was calculated for the negative hydrogen ion

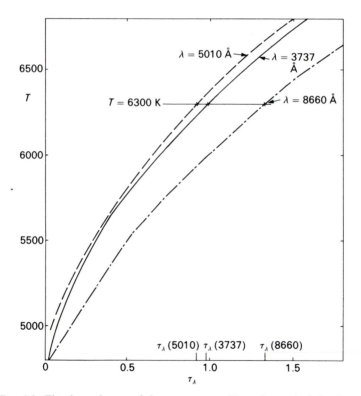

Fig. 6.3. The dependences of the temperature $T$ on the optical depth $\tau_{\lambda_1}, \tau_{\lambda_2}, \tau_{\lambda_3}$, where $\lambda_1 = 5010$ Å, $\lambda_2 = 3737$ Å, $\lambda_3 = 8660$ Å. The horizontal line connects points of equal temperature $T_0 = 6300$ K. These points must all belong to the same geometrical depth $t$. The corresponding optical depths for the different wavelengths can be read off at the abscissa.

H$^-$ (see Fig. 6.4). Rupert Wildt first suggested in 1938 that H$^-$ might be most important for the continuous absorption coefficient in the solar photosphere. It is still very difficult to measure the absorption coefficient for H$^-$ in the laboratory because there are so few H$^-$ ions. The solar photosphere needs a depth of roughly 100 km to reach an optical depth of 1. In the laboratory we can use higher densities, but the required path lengths are still very large. Therefore, it was very important when Chalonge and Kourganoff (1946) showed that the wavelength dependence of $\kappa_\lambda$ could be measured for the sun and that it did agree with the one calculated for H$^-$.

## 6.3    Radiative equilibrium

We pointed out above that the energy transport in the star, which has to bring the energy from the interior to the surface, requires that the temperature decreases from the inside out, because a thermal energy flow goes only in the direction of decreasing temperature. This discussion indicates that the energy flux through the star is related to the temperature gradient. Here we will show that we can actually calculate the temperature gradient if we know how much energy is transported outwards at any given point and by which transport mechanism.

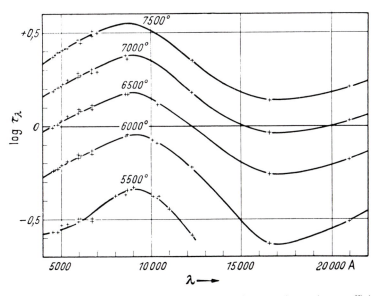

*Fig. 6.4.* The wavelength dependence of the continuous absorption coefficient $\kappa_\lambda$ in the solar photosphere as determined by Chalonge and Kourganoff in 1946. (From Unsöld, 1955, p. 116.)

We know from observation how much energy leaves the surface of a star per $cm^2$ each second. We expect that a larger temperature gradient is needed for a larger energy transport. We also expect a larger temperature gradient if the heat transport is made difficult by some obstacle. For instance, if we think about a room with an outside wall where the heat is leaving, and with a radiator on the inside wall on the opposite side, then there will be a temperature gradient from the radiator wall to the outside wall. The temperature gradient will be larger if the window is opened on a cool day, which will cause a large energy flux to go out the window. If a large flux is to go out the window the temperature gradient will have to be larger if shelves are erected in front of the radiator which inhibit the heat flow. The heat flow is largest for the largest temperature gradient and for the smallest number of obstacles. We can turn this around and say if a large amount of heat energy transport is required we need a steep temperature gradient, and if the energy transport is made difficult, we also need a steep temperature gradient in order to transport a given amount of energy.

If we have measured the amount of energy leaving the stellar surface, how do we know how much energy is transported outwards at a deeper layer? We know that the radiation we see comes from a layer which has a geometrical height of a few hundred kilometers. We can easily calculate how much thermal energy is contained in this layer. For instance, for the sun the number of particles per $cm^3$ is about $10^{17}$. The total number of particles in a column of 100 km height and a cross-section of $1\,cm^2$ is then roughly $10^{24}$. Each particle has a thermal energy of $\frac{3}{2}kT$, which is about $10^{-12}$ erg. The total thermal energy of this column is therefore about $10^{12}\,erg\,cm^{-2}$. The radiative energy loss per $cm^2$ of the solar surface is equal to $\pi F$, which for the sun is measured to be $6.3 \times 10^{10}\,erg\,cm^{-2}\,s^{-1}$. The energy content of the solar photosphere therefore can only last for 15 seconds if the sun keeps shining at a constant rate. However, we see that the photosphere is not cooling off; therefore, we know that the amount of energy which the sun is losing at the surface must constantly be put into the photosphere at the bottom. If exactly the same amount of energy were not replenished at any given time the solar photosphere would very rapidly change its temperature. If too little energy were to flow in from the bottom the photosphere would cool off; if too much were transported into the bottom of the photosphere it would heat up and more energy would leave the solar surface. Nothing like this happens, so the amount of energy flowing into the photosphere per $cm^2$ each second must equal exactly the amount of energy lost at the surface, which is $\pi F$. The same argument can be made for the layer below the photosphere. It too must get the same amount of energy from the next deeper layer in order to remain at the same temperature. Therefore, we must have the situation shown in Fig. 6.5. **The energy flux must remain constant with depth.**

This means

$$\frac{dF}{dt} = 0, \quad \text{or} \quad \frac{dF}{d\tau} = 0$$

or

$$\pi F = \text{const.} = \sigma T_{\text{eff}}^4. \qquad 6.9$$

Equation (6.9) is a basic condition for a star if it is to remain constant in time and not supposed to heat up or cool off. We call this the *condition of thermal equilibrium*. This should not be confused with the condition of thermodynamic equilibrium, for which we require that the temperature should be the same everywhere and that we should always find the same value of the temperature, no matter how we measure it. Nothing like this is required for what we call thermal equilibrium. **For thermal equilibrium we require only that the temperature should not change in time.**

So far we have not yet specified by which means the energy is transported. The condition of thermal equilibrium does not specify this. We can have different transport mechanisms at different layers. We know that at the surface the transport must be by radiation. If in the deeper layers the energy transport is also due to radiation, i.e., **if the heat flux $F$ is radiative heat flux, then we talk about** *radiative equilibrium*. As we said, this does not have to be the case. We may have some energy transport by mass motion – the so-called convective energy transport – or we might have energy transport by heat conduction. If all the heat transport were by convection, which, strictly speaking, never happens, we might talk about convective equilibrium. Also, if all the energy transport were due to heat conduction, we might talk about conductive equilibrium. The latter two

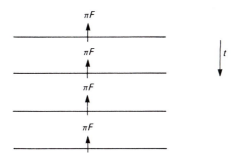

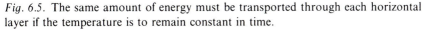

*Fig. 6.5.* The same amount of energy must be transported through each horizontal layer if the temperature is to remain constant in time.

cases are very rare in stars, so these concepts are hardly ever used. The scheme in Fig. 6.6 demonstrates the situation.

Generally, in the stellar photospheres we find that the condition of radiative equilibrium is a very good approximation. In the solar photosphere we see granulation showing that there is also energy transport by mass motion, but the convective flux is of the order of 1% of the total flux. If there is a temperature gradient there is also some conductive flux, but again, this is much smaller than both the radiative energy flux and the convective flux. When we know how large the temperature gradient is in the solar photosphere and in other stellar atmospheres, we will come back to the discussion of this question. In the following chapters we will only be concerned with radiative equilibrium.

For radiative equilibrium we now consider the energy flux $F = F_r$ to be radiative flux only. Equation (6.9) does not, of course, imply that the same photons keep flowing across the different horizontal planes. There is much absorption and re-emission, but the surplus of radiation in the outward direction, i.e., the net flux $F(\tau)$, has to remain constant.

In order to see what this means for the source function, we have to start from the transfer equation (5.3) which we integrate over the whole solid angle and obtain

$$\int \cos \vartheta \, \frac{\mathrm{d}I_\lambda(\tau_\lambda, \vartheta)}{\mathrm{d}\tau_\lambda} \cdot \mathrm{d}\omega = + \int I_\lambda(\tau_\lambda, \vartheta) \, \mathrm{d}\omega - \int S_\lambda(\tau_\lambda) \, \mathrm{d}\omega. \qquad 6.10$$

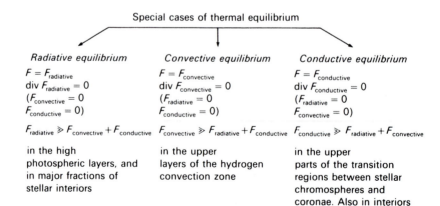

*Thermodynamic equilibrium*: nothing changes in time *and* space

*Thermal equilibrium*: no temperature changes in time
requires div $F = 0$
with $F = F_{\text{radiative}} + F_{\text{convective}} + F_{\text{conductive}}$

Special cases of thermal equilibrium

| *Radiative equilibrium* | *Convective equilibrium* | *Conductive equilibrium* |
|---|---|---|
| $F = F_{\text{radiative}}$ | $F = F_{\text{convective}}$ | $F = F_{\text{conductive}}$ |
| div $F_{\text{radiative}} = 0$ | div $F_{\text{convective}} = 0$ | div $F_{\text{conductive}} = 0$ |
| $(F_{\text{convective}} = 0$ | $(F_{\text{radiative}} = 0$ | $(F_{\text{radiative}} = 0$ |
| $F_{\text{conductive}} = 0)$ | $F_{\text{conductive}} = 0)$ | $F_{\text{convective}} = 0)$ |
| $F_{\text{radiative}} \gg F_{\text{convective}} + F_{\text{conductive}}$ | $F_{\text{convective}} \gg F_{\text{radiative}} + F_{\text{conductive}}$ | $F_{\text{conductive}} \gg F_{\text{radiative}} + F_{\text{convective}}$ |
| in the high photospheric layers, and in major fractions of stellar interiors | in the upper layers of the hydrogen convection zone | in the upper parts of the transition regions between stellar chromospheres and coronae. Also in interiors of white dwarfs. |

*Fig. 6.6* explains the concepts of the different equilibria.

Interchanging differentiation and integration on the left-hand side yields

$$\frac{d}{d\tau_\lambda}(\pi F_\lambda(\tau_\lambda)) = \int(-S_\lambda(\tau_\lambda) + I_\lambda(\tau_\lambda, \vartheta))\, d\omega. \qquad 6.11$$

Integrating the right-hand side and dividing by $4\pi$ gives

$$\frac{1}{4}\frac{d}{d\tau_\lambda} F_\lambda(\tau_\lambda) = -S_\lambda(\tau_\lambda) + J_\lambda(\tau_\lambda), \qquad 6.12$$

since $S_\lambda$ can be assumed to be isotropic and $J_\lambda = \int I_\lambda\, d\omega/4\pi$, according to (5.32).

If we now again assume $\kappa_\lambda$ to be independent of $\lambda$, i.e., we assume a *grey atmosphere*, we can then integrate over $\lambda$ and obtain

$$\frac{1}{4}\frac{d}{d\tau}\int_0^\infty F_\lambda(\tau)\, d\lambda = -\int_0^\infty S_\lambda(\tau)\, d\lambda + \int_0^\infty J_\lambda(\tau)\, d\lambda$$

or                                                                                  6.13

$$\frac{1}{4}\frac{d}{d\tau} F(\tau) = -S(\tau) + J(\tau) = 0,$$

according to equation (6.9). Equation (6.13) tells us that

$$S(\tau) = J(\tau). \qquad 6.14$$

In a grey atmosphere, the source function must be equal to the mean intensity $J$.

If the atmosphere is not grey, it is better to take $\kappa_\lambda$ to the right-hand side of equation (6.12). Integration over $\lambda$ and division by $4\pi$ then yields

$$\frac{1}{4}\frac{dF(\tau_\lambda)}{dt} = -\int_0^\infty (\kappa_\lambda S_\lambda - \kappa_\lambda J_\lambda)\, d\lambda = 0, \qquad 6.15$$

or

$$\int_0^\infty (\kappa_\lambda S_\lambda - \kappa_\lambda J_\lambda)\, d\lambda = 0, \qquad 6.16$$

which must always hold if (6.9) holds for the radiative flux, i.e., if we have radiative equilibrium.

The two conditions $dF/dt = 0$ and $\int_0^\infty \kappa_\lambda S_\lambda\, d\lambda = \int_0^\infty \kappa_\lambda J_\lambda\, d\lambda$ are equivalent.

Equation (6.16) can easily be understood: since $\int_0^\infty \kappa_\lambda S_\lambda\, d\lambda$ describes the total amount of energy emitted and $\int_0^\infty \kappa_\lambda J_\lambda\, d\lambda$ describes the total amount of energy absorbed per unit volume, equation (6.16) only says that the amount of energy absorbed must equal the amount of energy re-emitted if no heating or cooling is taking place.

We can understand the functioning of radiative equilibrium by comparing the emitted radiation with water being emitted from one fountain and thrown into the next, where it is 'absorbed' and mixed with the other water and new water is re-emitted and thrown into the next one, etc. Fig. 6.7 illustrates the situation. If the water level is to remain constant, equal amounts of water have to be 'absorbed' and 're-emitted', but the 'absorbed' water is generally not the same as the re-emitted water.

## 6.4  The theoretical temperature stratification in a grey atmosphere in radiative equilibrium

### 6.4.1  *Qualitative discussion*

As we saw earlier, the temperature stratification is determined by the amount of energy which has to be transported through the atmosphere. We know from observation that the amount $\pi F = \sigma T_{eff}^4$ is leaving the surface per $cm^2$ each second. This is the amount which has to be transported through every horizontal layer of the atmosphere. The larger the amount of energy transport, the larger the temperature gradient has to be. This means that the temperature gradient is expected to grow with increasing $T_{eff}^4$. We also saw that the more difficult the energy transport, the steeper the temperature gradient has to be. For the photons the obstacles are the

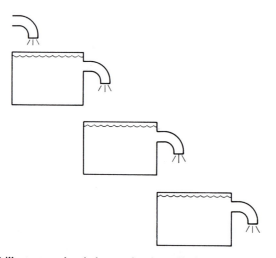

*Fig. 6.7* illustrates what is happening in radiative equilibrium. We compare the flow of photons through different horizontal layers with the flow of water through a series of fountains. The water (photons) is emitted from one fountain and flows into the next one. Here it is 'absorbed' and mixed with the rest of the water. A fraction of the mixed water is 're-emitted', i.e., it flows into the next fountain, is 'absorbed' there, mixed with the water (photons) in this fountain and another fraction is 'emitted' again and so on. If the water level (energy level) in all the fountains (horizontal layers of the photosphere) is required to remain constant, then the same amount of water (photons) has to flow out of each faucet into the next fountain in each unit of time.

atoms which keep absorbing them and emitting them equally in all directions. We therefore expect a steeper temperature gradient if the absorption coefficient $\kappa_\lambda$ is very large. $dT/dt$ is expected to grow with increasing $\kappa$. In fact, without any calculations we can estimate how fast the source function must grow with depth if the radiative flux is to remain constant. We can see this best from a series of drawings (see Fig. 6.8). Again, we compare the photosphere with a number of fountains. We know that on average the photons are absorbed after they have travelled a distance with an optical depth $\Delta\tau_s = 1$ in the direction of propagation. On average this corresponds to a vertical $\Delta\tau = \frac{2}{3}$, when averaged over all directions. We therefore give the fountains a size corresponding to an optical depth $\Delta\tau = \frac{2}{3}$ (see Fig. 6.8(a)). At the surface the flux $\pi F$ is flowing out. This amount of energy flowing into a half-sphere is indicated by *one* arrow. At the bottom of this layer the same amount – one arrow – has to be the *surplus* of energy flowing up as compared to the energy flowing down. The source function is determined by the photons emitted by the atoms. The atoms have no incentives to emit the photons preferentially into one direction. The source function is therefore expected to be isotropic. Since at the top just one arrow is flowing out, the emission, i.e., the source function, in the top panel must be one arrow in each direction. The source function in the top panel is completely determined by the outflow at the surface. So one arrow is flowing downward from the top fountains into the next deeper layer of fountains. Since the surplus flux upward has to be one arrow, there must be an upward flow of two arrows from the next deeper layer. Again, the source function is isotropic. Therefore, in the next deeper panel there must be an emission of two arrows in all directions. This means the source function must be twice as large as in the top panel. Following this construction to the next deeper layer of fountains, we see easily that for each absorption and re-emission process the emission into a half-sphere must increase by the amount $\pi F$ leaving the surface into a half-sphere. In Fig. 6.8(b) we consider the energy emitted into half-spheres and absorbed from half-spheres. The energy going into the half-sphere at the surface is given by $\pi F = \sigma T_{\text{eff}}^4$. So for each absorption and re-emission process, i.e., over $\Delta\tau = \frac{2}{3}$, the emission into the half-sphere, which is $2\pi S$, must increase by $\pi F$. This means

$$\frac{\Delta(2\pi S)}{\Delta\tau} = \frac{\pi F}{2/3} = \frac{3}{2}\pi F \quad \text{and} \quad \frac{\Delta S}{\Delta\tau} = \frac{3}{4}F. \qquad 6.17$$

Before we start the mathematical derivation, let us look at Fig. 6.8(a) once more and check what the relation is between $J$, the mean intensity, and $S$, the source function. The total emission is given by $4\pi S$, by all the outgoing arrows. The total absorption is given by all the ingoing arrows, which correspond to $4\pi J$. We see that, while we have an anisotropy in the radiation field because we always have more upward arrows than downward, we still have the same number of arrows leaving each cell as we have arrows coming in, telling us that, in spite of the anisotropy in the radiation field (determining $F$), we still have

$$S = J. \hspace{3cm} 6.18$$

### 6.4.2    *Mathematical derivation*

Let us now see how the depth dependence of the source function can also be obtained from the transfer equation. We multiply equation

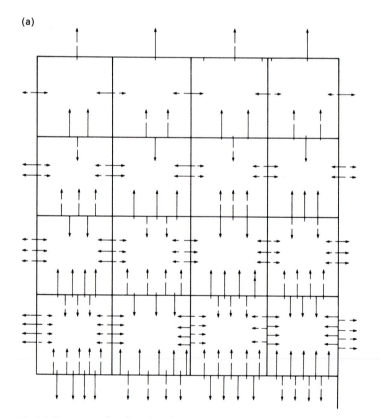

*Fig. 6.8.* (a) illustrates the situation in an atmosphere with radiative equilibrium. The source function must increase with depth if $F$ has to remain constant. (b) Here we plot the emissions into half-spheres for $\vartheta < \frac{1}{2}\pi$ and $\vartheta > \frac{1}{2}\pi$. The integrals over half-spheres have to grow by $\pi F$ for each step $\Delta\tau = \frac{2}{3}$.

(5.3) by cos $\vartheta$ and integrate over all solid angles, $\omega$, to obtain

$$4\pi \frac{d}{d\tau_\lambda} K_\lambda = \frac{d}{d\tau_\lambda} \int_{\omega=4\pi} \cos^2 \vartheta \, I_\lambda(\tau_\lambda) \, d\omega$$

$$= \int_{\omega=4\pi} \cos \vartheta \, I_\lambda(\tau_\lambda) \, d\omega - \int_{\omega=4\pi} \cos \vartheta \, S_\lambda(\tau_\lambda) \, d\omega. \qquad 6.19$$

The left-hand integral is generally abbreviated by $K_\lambda(\tau_\lambda)$. The last integral on the right-hand side is zero, since $S_\lambda(\tau)$ can be taken to be isotropic. The first integral on the right-hand side equals $\pi F_\lambda$. Division by $4\pi$ gives

$$\frac{dK_\lambda(\tau_\lambda)}{d\tau_\lambda} = \tfrac{1}{4} F_\lambda(\tau_\lambda). \qquad 6.20$$

We know that the temperature gradient is finally determined by the total energy transport through the atmosphere, i.e., by $\pi F$. For radiative

(b)

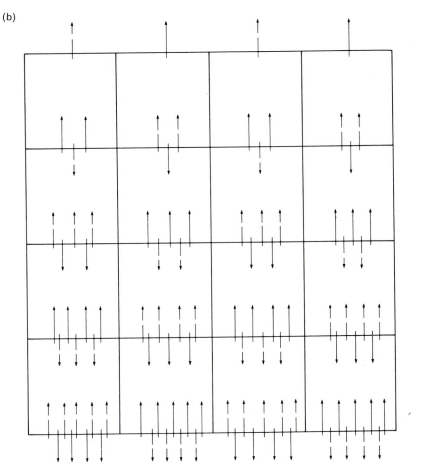

*Fig. 6.8(b)*

equilibrium, we therefore have to introduce the equation for radiative equilibrium, namely $dF/dt = 0$, into our equations. Since $F = \int_0^\infty F_\lambda \, d\lambda$, we have to integrate the equations over wavelengths $\lambda$. However, in equation (6.20) $\kappa_\lambda$ occurs in the denominator. If we want to integrate over $\lambda$ we have to specify the wavelength dependence of $\kappa_\lambda$. We have not yet discussed this, except for the empirical numerical determination for the sun; therefore, we will make the simplest specification; namely, assume that $\kappa$ is independent of wavelength, which means we now make the approximation of a *grey atmosphere*. Even if the real stellar atmospheres are not grey, we may hope that our results are still approximately correct if we determine an appropriate average value for $\kappa_\lambda$.

If we are dealing with a grey atmosphere we can integrate equation (6.20) over $\lambda$ and obtain

$$\frac{dK}{d\tau} = \tfrac{1}{4}F, \qquad \text{where } K = \int_0^\infty K_\lambda \, d\lambda. \qquad 6.21$$

This introduces a new unknown function $K(\tau)$. In order to find an equation for $dF/d\tau$ we now differentiate with respect to $\tau$, which, after integrating the transfer equation over $\lambda$, yields

$$\frac{d^2 K(\tau)}{d\tau^2} = \frac{1}{4}\frac{dF(\tau)}{d\tau} = J(\tau) - S(\tau) = 0, \qquad 6.22$$

because we are dealing with radiative equilibrium, for which $dF/d\tau = 0$. We then find again that for a grey atmosphere $S(\tau) = J(\tau)$. This is our first important result. Integration of the equation with respect to $\tau$ gives

$$K = c_1\tau + c_2, \qquad \text{where } \frac{dK}{d\tau} = c_1 = \tfrac{1}{4}F \text{ such that } K = \tfrac{1}{4}F\tau + c_2. \quad 6.23$$

With the new unknown function $K(\tau)$ we now have for a given $F$ two equations – (6.22) and (6.23) – for the determination of the three unknowns, $K, J, S$. We need an additional relation between two of these variables in order to determine all three.

We have seen that for the determination of the flux the anisotropy in the radiation field $I_\lambda(\vartheta)$ is very important because in the flux integral the inward-going intensities are subtracted from the outward-going ones, due to the factor $\cos\vartheta$. For $K$, on the other hand, a small anisotropy is unimportant because the intensities are multiplied by the factor $\cos^2\vartheta$, which does not change sign for inward and outward radiation. In order to evaluate $K$, we can therefore approximate the radiation field by an isotropic radiation field of the mean intensity $J$. From the definition of $K_\lambda$ we obtain

$$4\pi K_\lambda(\tau_\lambda) = \int_{\omega=4\pi} I_\lambda(\tau_\lambda, \vartheta) \cos^2 \vartheta \, d\omega$$

$$= J_\lambda(\tau_\lambda) \int_{\omega=4\pi} \cos^2 \vartheta \, d\omega = \frac{4\pi}{3} J_\lambda(\tau_\lambda), \qquad 6.24$$

or, after division by $4\pi$,

$$K_\lambda = \tfrac{1}{3}J_\lambda \qquad \text{and} \qquad K = \tfrac{1}{3}J, \qquad 6.25$$

when we integrate over $\lambda$. This approximation for the $K$ function is called the *Eddington approximation*. It has received wide application and is therefore very important.

Inserting the relation (6.25) into equation (6.21) we find

$$\frac{dK(\tau)}{d\tau} = \tfrac{1}{3}\frac{dJ(\tau)}{d\tau} = \tfrac{1}{4}F(\tau) = \text{const.} \qquad 6.26$$

From (6.23) we then derive

$$\frac{dK(\tau)}{d\tau} = c_1 = \tfrac{1}{4}F \qquad \text{or} \qquad \frac{dJ(\tau)}{d(\tau)} = 3\frac{dK(\tau)}{d\tau} = \tfrac{3}{4}F, \qquad 6.27$$

which yields

$$J(\tau) = \tfrac{3}{4}F\tau + \text{const.} = S(\tau). \qquad 6.28$$

Instead of equation (6.17), which we derived from Fig. 6.8(b), we now find again that

$$\frac{dS}{d\tau} = \tfrac{3}{4}F \qquad \text{and} \qquad S(\tau) = \tfrac{3}{4}F\tau + c_2, \qquad 6.29$$

as we derived from Fig. 6.8(b) (see equation 6.17). *From the condition of radiative equilibrium we thus find the law for the depth dependence of the source function.*

If we want to derive the depth dependence of the temperature from this, we have to know what the relation is between the source function and the temperature. The condition of radiative equilibrium does not tell us this directly. Therefore, we now make the assumption of LTE, which means we now assume that the source function is given by the Planck function. We want to emphasize that the temperature which we obtain as a function of depth depends on this assumption. The depth dependence of the source function is independent of this assumption. With

$$S(\tau) = B(\tau) = \frac{\sigma}{\pi} T^4(\tau), \qquad 6.30$$

we then find

$$\sigma T^4(\tau) = \tfrac{3}{4}\pi F(\tau + \text{const.}) = \pi S(\tau), \qquad 6.31$$

and

$$T^4(\tau) = \tfrac{3}{4}T^4_{\text{eff}}(\tau + \text{const.}) \qquad \text{since } \pi F = \sigma T^4_{\text{eff}}. \qquad 6.32$$

The source function must increase linearly with the optical depth $\tau$ and, as we saw, the gradient must be proportional to the flux.

The constant $c_2$ in equation (6.23) is still undetermined. Instead of this constant we can determine the constant in equation (6.32). These are integration constants which have to be determined from the boundary conditions. Our boundary condition here simply states that there is no flux going into the star. We must have

$$I(0, \vartheta) = I^- = 0 \qquad \text{for } \tfrac{1}{2}\pi < \vartheta \leqslant \pi. \qquad\qquad 6.33$$

In order to be able to calculate $K$ and $J$ at the surface of the star we make the simplifying assumption that the outward-going intensity does not depend on the angle $\vartheta$. This means we assume that

$$I(0, \vartheta) = I^+ = \text{const.} \qquad \text{for } 0 < \vartheta \leqslant \tfrac{1}{2}\pi. \qquad\qquad 6.34$$

With this the integration over the whole solid angle reduces to an integration over the half-sphere which gives

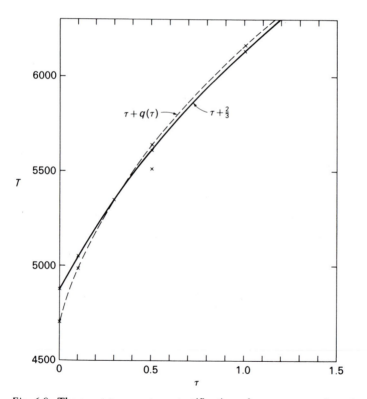

Fig. 6.9. The exact temperature stratification of a grey atmosphere (–––) is compared with the simple approximation (——) given by equation (6.38). The exact solution gives somewhat lower surface temperatures and slightly higher temperatures in deeper layers.

$$J(0) = \tfrac{1}{2}I^+ \qquad \text{and} \qquad F = I^+ \qquad \text{or} \qquad J(0) = \tfrac{1}{2}F, \qquad 6.35$$

which means

$$J(0) = B(0) = \frac{\sigma}{\pi}\, T^4(0) \qquad \text{or} \qquad T^4(0) = \tfrac{1}{2}T_{\text{eff}}^4. \qquad 6.36$$

From equation (6.32) we derive for $\tau = 0$

$$T^4(0) = \tfrac{3}{4}T_{\text{eff}}^4 \times \text{const.} = \tfrac{1}{2}T_{\text{eff}}^4, \qquad 6.37$$

which requires const. $= \tfrac{2}{3}$.

We finally derive for the grey atmosphere with the Eddington approximation

$$T^4(\tau) = \tfrac{3}{4}T_{\text{eff}}^4(\tau + \tfrac{2}{3}). \qquad 6.38$$

For the optical depth $\tau = \tfrac{2}{3}$ we then find $T^4(\tau = \tfrac{2}{3}) = T_{\text{eff}}^4$ or $T(\tau = \tfrac{2}{3}) = T_{\text{eff}}$, as we derived previously for a grey atmosphere.

Without making the Eddington approximation and with the accurate boundary condition, i.e., considering the true angular dependence of $I(0, \vartheta)$, one obtains

$$T^4(\tau) = \tfrac{3}{4}T_{\text{eff}}^4(\tau + q(\tau)), \qquad 6.39$$

where $q(\tau)$ is a function which varies slowly with $\tau$, with $q(0) = 0.577$ and $q(\infty) = 0.7101$. So $q(\tau) = \tfrac{2}{3}$ is quite good as an approximation. In Fig. 6.9 we compare the temperature stratifications obtained with (6.38) and (6.39).

How good an approximation is the grey atmosphere? In order to judge this, we have to discuss the frequency dependence of the absorption coefficients.

# 7

## The continuous absorption coefficient

### 7.1 The different absorption processes for hydrogen

Basically there are three different absorption mechanisms, two of which contribute to the continuous absorption, and one which gives rise to the line absorption. In order to understand this, let us look at the energy level diagram of the hydrogen atom, which we have plotted schematically in Fig. 7.1. The ionization energy for the hydrogen atom is 13.6 eV. This means that if an electron in the lowest energy level – the ground state – of the hydrogen atom in some way gains the energy of 13.6 eV, it has enough energy to overcome the attractive forces of the proton. It can then leave the atom and is a free electron. We have indicated the energy level which permits the electron to leave by the top level in Fig. 7.1. For the hydrogen atom the positions of the other energy levels can be calculated easily.

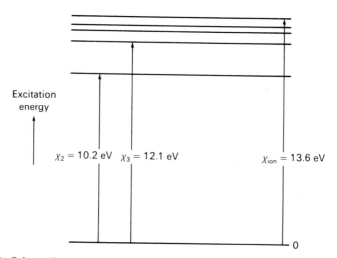

Fig. 7.1. Schematic energy level diagram for the hydrogen atom. Transitions between two discrete bound levels lead to line absorption or emission.

The difference between the first excited level ($n = 2$) and the ionization limit is, according to the Bohr model, 13.6/4 eV. The difference to the next higher one ($n = 3$) is $13.6/9 \approx 1.5$ eV. This yields for the excitation energy, i.e., the energy difference to the ground level,

$$\chi_n = \chi_{ion}\left(1 - \frac{1}{n^2}\right). \qquad 7.1$$

For the hydrogen atom, all orbitals of a given main quantum number $n$ but with different quantum numbers for angular momentum have the same energy. This means we can realize the same energy by several different sets of quantum numbers. The number of the different possibilities that will lead to the same energy level is called the *statistical weight* for that particular energy level. For a given $n$ this turns out to be $2n^2$. The statistical weight for each level $n$ in the hydrogen atom is therefore $2n^2$. We can realize a state with principal quantum number $n$ in $2n^2$ different ways. Since the Pauli principle does not permit two electrons with equal quantum numbers in one quantum cell, i.e., for one $n$, we find that we can have $2n^2$ electrons at each level $n$. This, of course, also applies to atoms with larger nuclear charges $Z$. This explains the shell structure of the atoms, which leads to noble gas configurations if there are $2n^2$ electrons in the outer shell. For atoms or ions with more than one electron, electron states with the same $n$ but with different other quantum numbers usually have different energy values.

A transition of an electron from one bound energy level to another leads to the emission or absorption of a photon, depending on whether the final energy level has a lower or higher energy than the starting level. The frequency of the emitted or absorbed photon is given by

$$h\nu = \chi_m - \chi_l, \qquad 7.2$$

if $\chi_m$ is the energy of the higher energy level and $\chi_l$ is that of the lower energy level involved in the transition. Since this gives a definite value for the frequency or the wavelength $\lambda = c/\nu$, a transition between two bound states can only lead to the emission or absorption of light of one particular wavelength, i.e., to line emission or absorption. We will talk about this later. We are presently concerned with the absorption between lines, i.e., an absorption mechanism that involves a continuum of frequencies. This is only possible if at least at one end of the transition there is a continuum of energy levels. We know that a free electron can have a continuous set of velocity values. If the transition begins or ends at a free state of the electron (i.e., an energy above the ionization limit) we can have a continuum of transition energies, which means a continuum of absorbed and emitted frequencies or wavelengths (see Fig. 7.2). We have basically two possibilities for continuous absorption. One possibility is a transition from a bound state at level $n$, to a free state with a velocity $v$. These transitions are called *bound-free transitions*. The frequency of the absorbed photon is given by

$$hv = (\chi_{ion} - \chi_n) + \tfrac{1}{2}mv^2. \qquad\qquad 7.3$$

We can also get a continuum of transitions if the electron goes from one free state with velocity $v_1$ to another free state with velocity $v_2$. The absorbed frequency for these *free-free transitions* is given by

$$hv = \tfrac{1}{2}mv_2^2 - \tfrac{1}{2}mv_1^2. \qquad\qquad 7.4$$

There are an infinite number of free-free transitions that will absorb the same frequency for different starting energies $\tfrac{1}{2}mv_1^2$. This holds for every frequency. For the calculation of the free-free absorption coefficient of a given atom or ion an infinite number of transition probabilities has to be computed at each frequency. For each discrete energy level $n$ there is only one bound-free transition which leads to an absorption of frequency $v$ in the continuum. However, for each frequency the transition probability has to be computed for each level $n$.

Each bound-free transition corresponds to an ionization process, since the electron is free afterwards. The *emission* of a photon by a free-bound transition corresponds to a recombination process.

For the hydrogen atom, the continuous absorption coefficient for one hydrogen atom at level $n$ was computed to be

$$a_n = \frac{1}{n^2} \frac{64\pi^4}{3\sqrt{3}} \frac{1}{v^3} \frac{Z^4 m e^{10}}{ch^6 n^3} G, \qquad\qquad 7.5$$

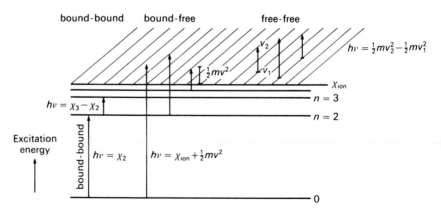

Fig. 7.2. A continuum of absorption processes can occur, if the electron transfers from a bound state into a free state, or if it goes from one free state into another one. We call these transitions bound-free or free-free transitions.

where $G$ is a correction factor, the so-called Gaunt factor, which does not deviate from 1 by more than a few percent. Here $m$ and $e$ are electronic mass and charge. $Ze$ is the effective nuclear charge attracting the absorbing electron.

In order to compute the total absorption at frequency $\nu$, we have to multiply $a_n$ by the number of atoms in the considered state with quantum number $n$ and then sum up over all states $n$ that contribute at the given frequency $\nu$. Finally, we have to add the free-free contributions.

Which states $n$ contribute at a given frequency? From Fig. 7.2 it is obvious that the minimum photon frequency necessary for a transition of an electron into the continuum – or for an ionization process – from level $n$ is given by $\chi_{ion} - \chi_n$. For the hydrogen atom $\chi_n = \chi_{ion}(1 - (1/n^2))$. For continuous absorption from level $n$ we therefore need a photon with an energy large enough to overcome this energy difference, i.e.,

$$h\nu > \chi_{ion} - \chi_n \quad \text{or} \quad \lambda < \frac{hc}{(\chi_{ion} - \chi_n)}. \qquad 7.6$$

For very long wavelengths only energy levels with very large $\chi_n$ can contribute to the continuous absorption. For very large $\lambda$ mainly free-free transitions will contribute. The contribution of level $n$ will start at the frequency $\nu_n = (\chi_{ion} - \chi_n)/h$ or $\lambda_n = hc/(\chi_{ion} - \chi_n)$ and will continue for shorter $\lambda$. At these frequencies the $\kappa_\lambda$ increases discontinuously because the number of absorbing atoms suddenly increases. The wavelength dependence of the continuous $\kappa$ for the hydrogen atom is shown in Fig. 7.3. In Fig. 7.4 we show the qualitative run of the $\mathrm{He}^+$ continuous absorption coefficient. For this ion the ionization energy is larger by a factor of $Z^2 = 4$ than that of the hydrogen atom. All excitation energies are larger by the same factor, and all discontinuities occur at wavelengths shorter by that same factor of 4. In both plots the quantum mechanical correction factors – the so-called Gaunt factors – were omitted. They change the $\kappa$ by a few percent only. In accurate calculations they have to be included.

We see in Fig. 7.3 that in the visual spectral region it is mainly the Paschen continuum, i.e., absorption from the level $n = 3$, which determines the hydrogen absorption coefficient. In the ultraviolet it is the Balmer continuum, i.e., absorption from the level $n = 2$. The discontinuity at $\lambda = 3647$ Å is determined by the ratio of the hydrogen atoms in the second and third quantum levels.

Another interesting fact is seen in Figs. 7.3 and 7.4; namely, the absorption coefficient per atom at the Lyman limit always has approximately the value of the cross-section of the lowest orbital in the atom, which for the hydrogen atom is $0.5 \times 10^{-8}$ cm. The cross-section $\sigma$ is therefore $\sigma \sim 0.785 \times 10^{-16}$ or $\log \sigma = 0.895 - 17$, which is nearly equal to $\log \kappa_\lambda$ at the Lyman limit as seen in Fig. 7.3. For the $He^+$ ion the radius of the orbital is smaller by the factor $Z$. The cross-section is then smaller by a factor of 4, as is the case for $\kappa_\lambda$ at the Lyman limit for the $He^+$ ion.

## 7.2    The Boltzmann formula

In order to calculate the absorption coefficient for the hydrogen atom $\kappa(H)$, we must know the number of hydrogen atoms in the different quantum states. We call these the *occupation numbers* of the different energy levels with excitation energies $\chi_n$ above the ground level.

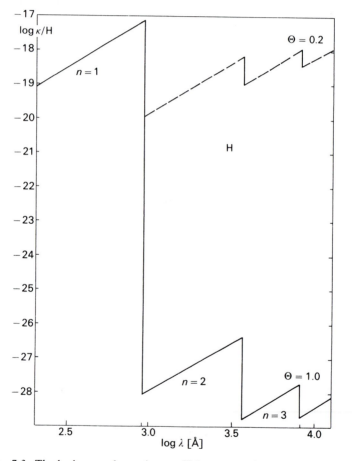

*Fig. 7.3.* The hydrogen absorption coefficient $\kappa_\lambda$ per hydrogen atom is shown as a function of wavelength for two temperatures $\Theta = 5040/T = 0.2$ and for $\Theta = 5040/T = 1.0$. Higher temperatures lead to higher values of $\kappa_\lambda$ in the visual spectral region.

Boltzmann showed that the probability of finding an atom or electron in an excited state with energy $\chi_n$ decreases exponentially with increasing $\chi_n$ but increases with increasing temperature $T$, i.e.,

$$\frac{N_n}{N_1} = \frac{g_n}{g_1} e^{-\chi_n/kT}, \qquad 7.7$$

where $\chi_n$ is the energy difference between the level $n$ and the ground level, $g_n$ and $g_1$ are the statistical weights for level $n$ and the ground state, respectively. Equation (7.7) is called the *Boltzmann formula*; $k$ is the Boltzmann constant, $k = 1.38 \times 10^{-16}$ erg degree$^{-1}$.

The Boltzmann formula in equation (7.7) is only a special form of the general case that the number of particles with an energy $E$ is always proportional to $e^{-E/kT}$ (see, for instance, the Maxwell distribution for the kinetic energies of particles).

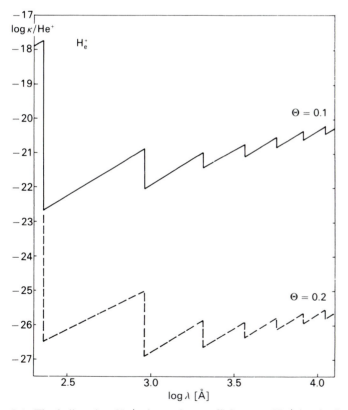

*Fig. 7.4.* The helium ion He$^+$ absorption coefficient per He$^+$ ion is shown as a function of wavelength $\lambda$ for the two temperatures $\Theta = 5040/T = 0.1$ and $\Theta = 5040/T = 0.2$. The He$^+$ absorption coefficient is extremely small for temperatures below 25 000 K.

In the exponent in equation (7.7) we find $k$, which is normally given in erg per degree, while the excitation energy is usually given in electron-volts. We would therefore like to rewrite the exponent in the form that we can use these quantities in the customary units. With $1\,\mathrm{eV} = 1.602 \times 10^{-12}$ erg we can rewrite the exponent in the following form:

$$\chi_n/kT = \frac{\chi_n(\mathrm{eV}) \times 1.602 \times 10^{-12}}{1.38 \times 10^{-16} \times T} = \frac{\chi_n(\mathrm{eV}) \times 1.160 \times 10^4}{T},$$

and

$$\log e^{-\chi_n/kT} = -(\chi_n/kT)\log e$$

$$= \frac{-\chi_n(\mathrm{eV})}{T} \times 1.160 \times 10^4 \times 0.4343 = -\chi_n(\mathrm{eV})\frac{5040}{T}. \qquad 7.8$$

The ratio $5040/T$ occurs in many places. It is usually abbreviated by the symbol $\Theta$.

In order to compute $N_n$ according to equation (7.7), we have to know $N_1$, which requires a knowledge of the abundances of the element under consideration. We also have to know whether the element is partially or almost completely ionized. We saw above that an electron bound in the atom may absorb a photon. If the photon is energetic enough the electron can get enough energy to leave the atom as a free electron and the former atom remains as an ion, i.e., as a particle with a positive charge. Such ionization decreases the number of atoms in the ground level. We therefore have to know how many atoms have been converted into ions.

## 7.3    The Saha equation

The degree of ionization of any atom or ion can be obtained from the Saha equation, which can be derived from the above Boltzmann formula if we extend the states with quantum numbers $n$ to positive energies, i.e., to free electrons with the appropriate statistical weights (see Unsöld, 1955). One finds

$$\frac{N_1^+ n_e}{N_1} = \frac{g_1^+}{g_1} 2 \frac{(2\pi mk)^{3/2}}{h^3} T^{3/2} e^{-\chi_{\mathrm{ion}}/kT}. \qquad 7.9$$

Here $m$ is the mass of the electron, $N_1^+$ is the number of ions in their ground state, and $N_1$ is the number of atoms in the ground state. $g_1^+$ is the statistical weight of the ground state of the ion.

Similarly, we can write down the Saha equation for the next higher state of ionization, if an additional electron is removed from the ion and an ion with two positive charges is formed. In this case we have to write

$$\frac{N_1^{++} n_e}{N_1^+} = \frac{g_1^{++}}{g_1^+} \frac{2(2\pi mk)^{3/2}}{h^3} T^{3/2} e^{-\chi_{\mathrm{ion}}^+/kT}, \qquad 7.10$$

where $g_1^{++}$ is the statistical weight of the ground state of the twice ionized ion and $N_1^{++}$ is the number of twice ionized ions in the ground state. $\chi_{ion}^+$ is the amount of energy necessary to remove the additional electron from the ion with one positive charge. It does not include the energy necessary to remove the first electron.

If we want the ratio of the total number of ions $N^+$ to the total number of atoms $N$, then we have to calculate $N/N_1$ and $N^+/N_1^+$. We can do this by using equation (7.7). Clearly, $N$ is the sum of all the particles in the different quantum states, $N_n$, and similarly, $N^+$ is the sum over all the ions in their different quantum states, $N_n^+$. With

$$N_n^+ = N_1^+ \frac{g_n^+}{g_1^+} e^{-\chi_n^+/kT},$$

and

$$N^+ = N_1^+ + \sum_{n=2}^{\infty} N_n^+ = N_1^+ + \frac{N_1^+}{g_1^+} \sum_{n=2}^{\infty} g_n^+ e^{-\chi_n^+/kT}, \qquad 7.11$$

we find

$$N^+ = \frac{N_1^+}{g_1^+} \left( g_1^+ + \sum_{n=2}^{\infty} g_n^+ e^{-\chi_n^+/kT} \right) = \frac{N_1^+}{g_1^+} u^+, \qquad \text{or} \qquad \frac{N^+}{N_1^+} = \frac{u^+}{g_1^+}, \qquad 7.12$$

where

$$u^+ = g_1^+ + \sum_{n=2}^{\infty} g_n^+ e^{-\chi_n^+/kT} \qquad 7.13$$

is called the *partition function* of the ion.

For the atom on the next lower state of ionization, we find similarly for its partition function $u$

$$u = g_1 + \sum_{n=2}^{\infty} g_n e^{-\chi_n/kT}. \qquad 7.14$$

Multiplication of (7.9) by $\dfrac{N^+}{N_1^+} \times \dfrac{N_1}{N}$ yields

$$\frac{N^+ n_e}{N} = \frac{u^+}{u} \frac{2(2\pi m)^{3/2}}{h^3} (kT)^{3/2} e^{-\chi_{ion}/kT}. \qquad 7.15$$

This is the well-known *Saha equation*. It is the equivalent of the mass reaction equation used in chemistry to compute the concentrations of different reacting particles.

For the next higher degree of ionization, we obtain

$$\frac{N^{++} n_e}{N^+} = \frac{u^{++}}{u^+} \frac{2(2\pi m)^{3/2}(kT)^{3/2}}{h^3} e^{-\chi_{ion}^+/kT}, \qquad 7.16$$

with $\chi_{ion}^+$ the ionization energy of $N^+$ to make $N^{++}$. Collecting all the atomic constants and writing equation (7.16) in logarithmic form, we obtain

$$\log \frac{N^+}{N} = \log \frac{u^+}{u} + \log 2 + \tfrac{5}{2} \log T - \chi_{\text{ion}} \Theta - \log P_e - 0.48. \qquad 7.17$$

Here we have introduced $P_e = n_e kT$, the electron pressure, instead of the electron density $n_e$. We therefore had to add $\log k + \log T$.

In this logarithmic form the Saha equation contains only one constant which can easily be remembered.

We are now able to calculate the degree of ionization for the hydrogen atom, for instance, in the solar photosphere, where we have a temperature of $T \approx 6000$ K and where the electron pressure is $\log P_e = 1.5$, as we shall see later. For this temperature we obtain $\Theta = 0.84$, $\log T = 3.78$ and $2.5 \log T = 9.45$. Since the proton has no electrons left to be in excited states, it can only exist in one form. Its statistical weight for the ground state is 1 and the partition function is also 1. The neutral hydrogen atom can exist in two forms in the ground level; with spin-up or with spin-down for the electron. The statistical weight of the ground level is therefore 2. The excitation energies for the levels with $n > 1$ are so high that the numbers of particles in those states are negligible in comparison with the number in the ground state. We can therefore say that the partition function $u(\text{H}) = 2$. We then obtain the Saha equation for the ionization of hydrogen in the solar photosphere

$$\log \frac{\text{H}^+}{\text{H}} = \log \tfrac{2}{2} + 9.45 - 0.48 - 11.42 - 1.5 = -4,$$

or $\log(N(\text{H}^+)/N(\text{H})) = -4$, which tells us that $\text{H}^+/\text{H} = 10^{-4}$, i.e., only one out of $10^4$ hydrogen atoms is ionized.

Let us now do the same computation for the negative hydrogen ion $\text{H}^-$. The binding energy for the electron in this ion is only $\chi_{\text{ion}}(\text{H}^-) = 0.7$ eV. The $\text{H}^-$ ion has no excited states with energies below the ionization energy. We only have to consider the statistical weight for the ground level if we want to calculate the partition function. The $\text{H}^-$ ion has two electrons; they are both in the level with $n = 1$. Since, according to the Pauli principle, these two electrons cannot have all quantum numbers equal, one electron must have spin-up and the other must have spin-down. There is no other way for them to exist. This means that the statistical weight for the $\text{H}^-$ ion in the ground state must be 1. We also have to realize now that it is the $\text{H}^-$ ion which loses the electron to become a neutral atom. Therefore, the hydrogen atom now is in the role of the ion, while the $\text{H}^-$ is in the place of the neutral atom. The Saha equation therefore has to be written as

$$\log \frac{N(\text{H})}{N(\text{H}^-)} = \log \tfrac{4}{1} + 9.45 - 0.48 - 0.59 - 1.5 = +7.5,$$

or

$$\frac{N(H^-)}{N(H)} \approx 3 \times 10^{-8}.$$

Only one out of $10^7$ hydrogen atoms is in the form of $H^-$.

## 7.4 The $H^-$ absorption coefficient in the sun

If there are so few $H^-$ ions in the solar photosphere, why then is the $H^-$ absorption coefficient so important?

We saw earlier that in the visual spectral region only the hydrogen atoms in the third quantum level can contribute to the continuum absorption, i.e. in the Paschen continuum. For the $H^-$ ion with an ionization energy of only 0.7 eV, the limiting wavelengths for continuum bound-free absorption is at about 17 000 Å. This means all the $H^-$ ions can contribute to the absorption in the visual region. In order to estimate the importance of the $H^-$ absorption we therefore have to compare the number of $H^-$ ions with the number of hydrogen atoms in the third quantum level. From the Boltzmann formula we find that

$$\log \frac{N_H(n=3)}{N_H(n=1)} = 0.95 - 12.1\Theta = 0.95 - 12.1 \times 0.84 = 0.95 - 10.16 = -9.21.$$

or

$$\frac{N_H(n=3)}{N_H(n=1)} = 6 \times 10^{-10}.$$

Since

$$\frac{N(H^-)}{N_H(n=3)} = \frac{N(H^-)}{N_H(n=1)} \frac{N_H(n=1)}{N_H(n=3)},$$

we find

$$\frac{N_H(n=3)}{N(H^-)} = \frac{6 \times 10^{-10}}{3 \times 10^{-8}} = 2 \times 10^{-2}; \text{ twice.}$$

As the atomic absorption coefficients per absorbing atom are generally of the same order of magnitude, we expect the $H^-$ absorption to be roughly 100 times more important than the Paschen continuum.

On the other hand, if we compute the number of atoms in the second quantum level, we find

$$\log \frac{N(n=2)}{N(n=1)} = 0.6 - 10.2\Theta = -7.97 \sim -8,$$

and

$$\frac{N(n=2)}{N(H^-)} = \frac{10^{-8}}{10^{-7}} \sim 10^{-1}.$$

There are 10 times as many H$^-$ ions as hydrogen atoms in the second quantum level, but the number of hydrogen atoms in the second quantum level is not completely negligible. At the Balmer limit, i.e., at $\lambda$3647 Å, the continuous $\kappa$ in the sun is expected to increase by about 10%. There is a small discontinuity observed in the continuous $\kappa$, as seen in Fig. 7.5.

What is the wavelength dependence of the H$^-$ absorption coefficient? We have one bound-free continuum from the ground level of H$^-$. There are no excited levels in H$^-$. The bound-free continuous absorption can take place for all $\lambda$ shorter than that corresponding to $h\nu = \chi_{ion} = 0.7$ eV. This corresponds to a wavelength of 17 000 Å. For longer wavelengths, we can only have free-free absorption, which goes roughly $\sim \nu^{-3}$.

The bound-free $\kappa(H^-)$ shows a different behavior from $\kappa(H)$. It first increases when going to shorter wavelengths, has a maximum around 8000 Å, and then decreases towards the violet. The sum of the free-free and the bound-free absorption shows a minimum around 16 000 to

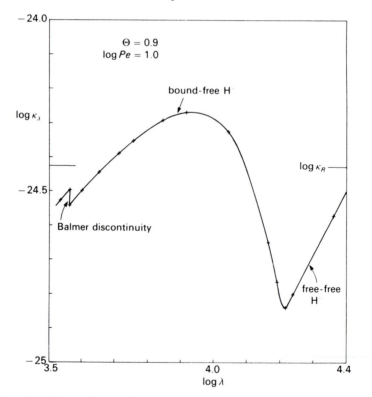

*Fig. 7.5.* The continuous absorption coefficient per heavy particle is shown for $\Theta = 0.9$, which means $T = 5600$, and $\log P_e = 1.0$. The contributions from the bound-free and free-free continua of H$^-$ are dominant. The small contribution from the Balmer continuum of hydrogen is visible at $\log \lambda = 3.562$. $\bar{\kappa}_R$ is the Rosseland mean absorption coefficient; see section 8.4.

16 500 Å, depending on the temperature and the relative importance of the free-free and bound-free absorption.

We have seen earlier how we can test the frequency dependence of the absorption coefficient in the sun by means of the center-to-limb variation of the intensities in different frequencies (see Section 6.2). We can now see that the measured and the theoretical wavelength dependences do indeed agree quite well.

## 7.5    The helium absorption in the sun

Helium is the most abundant element after hydrogen. Could the helium atom be important for the continuous absorption in the sun or in other stars? Helium has an ionization energy of about 24 eV. The first excited states have excitation energies of about 20 eV. For solar temperatures the number of atoms in the first excited state can again be estimated from the Boltzmann formula

$$\log \frac{N_{\mathrm{He}}(n=2)}{N_{\mathrm{He}}(n=1)} \sim 0.9 - 20.0\theta \approx -16.$$

Only a fraction of $10^{-16}$ of the He atoms can contribute to the absorption. Since helium is about 10% as abundant as hydrogen, only one out of $10^{17}$ atoms are helium atoms in the second quantum level. The helium absorption from the ground level can only occur for wavelengths shorter than 500 Å, where there is very little radiation coming from the sun.

## 7.6    The metallic absorption in the sun

Let us now look at the metals, such as iron. Its abundance is roughly $10^{-4}$ times that of hydrogen. The ionization energy is 7.9 eV. Let us compute the number of iron atoms which can absorb at 4000 Å. 4000 Å corresponds to an energy $h\nu$ of 3.1 eV. Fig. 7.6 shows that the atoms have to be in a state which has an excitation energy greater than 4.8 eV if they are to absorb at 4000 Å. The iron atom has many energy levels. We can therefore assume that it has one at this particular energy. If we assume that the statistical weights of the different levels are all equal, we can get a rough estimate for the number of atoms in the state with an excitation energy of 4.8 eV by again applying the Boltzmann formula

$$\log \frac{\mathrm{Fe}(\chi_n)}{\mathrm{Fe}(\chi=0)} = -4.8 \times 0.84 = -4.03 \quad \text{or} \quad \frac{\mathrm{Fe}(\chi_n)}{\mathrm{Fe}(\chi=0)} = 10^{-4}.$$

With an iron abundance $N(Fe)/N(H) \approx 10^{-4}$, it seems that the number of neutral Fe atoms absorbing at 4000 Å could be comparable with the number of H$^-$ ions. However, we also have to consider the degree of ionization of iron, which can be obtained from the Saha equation.

$$\log \frac{Fe^+}{Fe} = -0.48 + \tfrac{5}{2} \log T - 7.9\Theta - \log P_e$$

$$= -0.48 + 9.45 - 6.64 - 1.5 = 0.8,$$

where we have assumed $2u^+/u \approx 1$, which is not quite correct, but accurate enough for this order of magnitude estimate.

In the solar atmosphere, iron is mainly ionized. This reduces the number of neutral iron atoms by roughly a factor of 10. Therefore, at 4000 Å the metals are not important, but let us make the same kind of estimate for a wavelength of 2000 Å. This wavelength corresponds to an energy of 6.2 eV. Therefore, iron atoms only need an excitation energy of $7.9 - 6.2 \text{ eV} = 1.7 \text{ eV}$. According to the Boltzmann formula, we find for this excitation energy a fraction of excited Fe atoms of

$$\log \frac{Fe(\chi_n)}{Fe(\chi = 0)} = -1.7 \times 0.84 = -1.43 \quad \text{or} \quad \frac{Fe(\chi_n)}{Fe(\chi = 0)} = 3.7 \times 10^{-2}.$$

Considering the degree of ionization and multiplying this with the abundance ratio of Fe, we obtain the fraction of Fe atoms absorbing at 2000 Å relative to the number of neutral hydrogen atoms to be about $6 \times 10^{-7}$, while the ratio of H$^-$ ions to the number of neutral hydrogen atoms is $3 \times 10^{-8}$, as we saw in Section 7.3. For the absorption at 2000 Å, we have roughly 20 times as many Fe atoms in the appropriate state of excitation than we have H$^-$ ions. At these short wavelengths the metallic atoms are therefore much more important for the absorption than the H$^-$ ion or the neutral hydrogen atom.

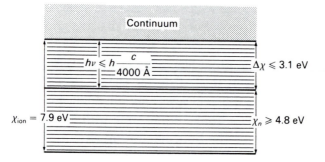

*Fig. 7.6.* Schematic energy level diagram for iron. All iron atoms with excitation energies $\chi_n > 4.8$ eV can contribute to the bound-free transitions for $\lambda < 4000$ Å.

Even more important is the absorption by the metal atoms in the ground level which for the iron atoms occurs for wavelengths shorter than 1570 Å, and for the Si atoms with an ionization energy of 8.15 eV at a wavelength of 1520 Å. We find a very pronounced discontinuity in the overall absorption coefficient at these wavelengths which leads to very pronounced discontinuities in the observed energy distributions of A, F and G stars, as we shall soon see.

## 7.7     Scattering by atoms and ions

In the classical picture of an atom we can consider the electron as being bound to the atom. Any force trying to remove it will be counteracted by an opposing force trying to bring it back into its equilibrium position. The situation is similar to the picture of an electron bound to the nucleus by some sort of spring (see Fig. 7.7). Such a spring would start to oscillate if a force were to try to pull on the electron and then let go. It would oscillate with certain eigenfrequencies $\omega_r = 2\pi v_r$, which are determined by the restoring force. The stiffer the spring, the more rapid the oscillations will be. In the case of the atom, the oscillation frequencies are determined by the Coulomb forces of the nucleus. For the hydrogen atom in the ground state the frequencies of the Lyman lines are the eigenfrequencies. Going back again to the picture of the spring, we can see that if somebody tries to make it oscillate with a frequency $\omega$, the amplitudes would get larger the closer $\omega$ is to any of the resonance frequencies or eigenfrequencies $\omega_r$. We can calculate that the amplitude would be proportional to $(\omega_r^2 - \omega^2)^{-1}$ if there is no damping. We know that light consists of electromagnetic

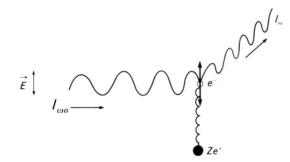

*Fig. 7.7.* The electron bound to the nucleus of an atom will start oscillating in the electric field $E$ of the lightwave with intensity $I_{\omega 0}$. This oscillation will cause it to radiate into all directions. The emission $I_\omega$ has the frequency $\omega$ of the forced oscillation, which means the emitted light has the same frequency as the original light beam. The light is scattered, but mainly in the directions perpendicular to the direction of oscillation.

waves, which means that in the presence of light the electron bound to the atom experiences an oscillating electric field which excites the bound electron to oscillations. The amplitude of these oscillations is also proportional to $(\omega_r^2 - \omega^2)^{-1}$. The electron in the atom has many eigenfrequencies $\omega_r$, namely, all the Lyman lines in the case of the hydrogen atom. All the resonance frequencies contribute to the amplitude of oscillation. The probability of each resonance frequency contributing is given by the so-called oscillator strength $f_r$, describing the probability for that particular wavelength to be absorbed or emitted by the atom. This oscillator strength $f_r$ is, of course, the same quantity which also determines the relative strengths of the Lyman lines.

If an electron is oscillating, it is accelerated. Each accelerated charge radiates; therefore, the oscillating electron also radiates. The radiated intensity is proportional to the square of the acceleration, which is $\ddot{x}$ if the displacement is called $x$. We describe the oscillating displacement $x$ by

$$x = A \cos \omega t, \qquad \text{where } \omega = 2\pi\nu,$$

then

$$\dot{x} = -\omega A \sin \omega t$$

which gives

$$\ddot{x} = -\omega^2 A \cos \omega t. \tag{7.18}$$

The radiated intensity is then proportional to

$$\ddot{x}^2 = x^2 \omega^4. \tag{7.19}$$

If damping is neglected, the emission corresponding to each eigenfrequency $\omega_r$ turns out to be

$$\text{Emission} = I_\omega \propto \frac{\omega^4}{(\omega_r^2 - \omega^2)^2}. \tag{7.20}$$

As we said earlier, for the neutral hydrogen atom the resonance frequencies correspond to the Lyman lines which are in the far ultraviolet. For the optical region we therefore find $\omega_r \gg \omega$ such that the scattering coefficient, also called the *Rayleigh scattering coefficient*, $\sigma_R$, is

$$\sigma_R \propto C \sum_r f_r \left( \frac{\omega^2}{\omega_r^2 - \omega^2} \right)^2 \approx \sum_r C \frac{f_r \omega^4}{\omega_r^4}, \tag{7.21}$$

if we again neglect the damping. We can replace this by writing

$$\sigma_R = C \frac{\omega^4}{\bar{\omega}_r^4}$$

with the definition

$$\frac{1}{\bar{\omega}_r^4} = \sum_r \frac{f_r}{\omega_r^4}, \qquad \text{where } C = \frac{8\pi}{3} \left( \frac{e^2}{mc^2} \right)^2, \tag{7.22}$$

and $m$ and $e$ are again electron mass and charge. Since for the hydrogen atom $\sum f_r = 1$ for all transitions from the ground level, one finds that $\bar{\omega}_r = \omega_0$ differs only little from $\omega$ for Lyman $\alpha$; it actually corresponds to 1026 Å. The Lyman $\alpha$ line has a wavelength of 1215 Å.

For the Rayleigh scattering coefficient $\sigma_R$ the final expression is

$$\sigma_R \text{ per atom} = C\left(\frac{\omega}{\omega_0}\right)^4 = C\left(\frac{\lambda_0}{\lambda}\right)^4 \propto \frac{1}{\lambda^4}, \qquad 7.23$$

i.e., $\sigma_R$ is steeply increasing towards the blue.

The light of our sky is scattered sunlight; scattered by the molecules in our atmosphere. From equation (7.23) we see that the scattering coefficient increases steeply towards shorter wavelengths. The amount of scattered light which we see in the sky also increases steeply towards shorter wavelengths, which means towards the blue. Therefore, the sky appears blue to us.

In the modern picture of the atoms, we would have to think about the Rayleigh scattering on hydrogen atoms as the sum of the overlapping Lyman line wings. However, only the pure scattering fraction is considered. Lyman $\alpha$ is the most important line.

## 7.8 Thomson scattering by free electrons

We still have to consider the electrons which are not bound to an atom. There is no restoring force; their eigenfrequencies are zero. We then find, according to equation (7.20), that

$$\sigma_{el} \propto C \frac{\omega^4}{\omega^4} = C, \qquad 7.24$$

which means the scattering coefficient is independent of wavelength. The value of $C$ is the same as in equation (7.22). It comes out to be

$$\sigma_{el} = \frac{8\pi}{3}\left(\frac{e^2}{mc^2}\right)^2 \qquad 7.25$$

per electron, or

$$\sigma_{el} = 0.66 \times 10^{-24} \text{ cm}^2.$$

This is the *Thomson scattering coefficient*. The Rayleigh scattering coefficient can then be written as

$$\sigma_R = \sigma_{el}\left(\frac{\omega}{\omega_0}\right)^4. \qquad 7.26$$

Quantum mechanical calculations give slightly lower values, especially if the frequencies are close to the resonance frequencies. We should have considered the damping due to the radiation.

Numerically, the electron scattering coefficient per free electron comes out to be $\sigma_{el} = 0.66 \times 10^{-24}$ cm$^2$. The absorption coefficient per neutral hydrogen atom at the Lyman limit is of the order $10^{-17}$ cm$^2$. The number of free electrons in the solar atmosphere as compared to the number of hydrogen atoms is given approximately by the ratio of the electron pressure to the gas pressure, which is $3 \times 10^{-3}$. Compared to the abundance of the H$^-$ ions the electrons are much more abundant, but since the scattering coefficient is so much lower than the absorption coefficient per atom or ion in the ground state, electron scattering is not important in the solar atmosphere. But this is different in stars where all the atoms are ionized. Considering the Rayleigh scattering by hydrogen atoms, we gain a factor $3 \times 10^3$ because the hydrogen atoms are so much more abundant in the sun than the electrons. At the shorter wavelengths, i.e., $\lambda \ll 3000$ Å, the contribution of Rayleigh scattering in the sun is not negligible, although the metallic absorption coefficients are generally larger than the Rayleigh scattering coefficient.

Altogether, we find absorption coefficients in the solar atmosphere as shown in Fig. 7.8, which closely resembles the wavelength dependence

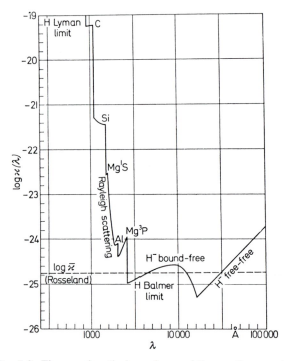

*Fig. 7.8.* The wavelength dependence of the continuous absorption coefficient per particle is shown for $T_{eff} = 5040$ K ($\Theta = 1.0$) and $\log P_e = 0.5$.

determined empirically by Chalonge and Kourganoff (1946) for the solar atmosphere.

## 7.9    Absorption coefficients for A and B stars

For solar temperatures we found that the $H^-$ ion contributes most to the continuous absorption because only hydrogen atoms in the third quantum level can absorb in the visual region. For solar temperatures the number of hydrogen atoms in the third quantum level is very small. For stars with higher temperatures this situation may well change.

Let us look now at the situation in an A star with $T \approx 10\,000$ K, $\log P_e = 3.0$, $\log T = 4.0$, $\Theta = 0.5$.

For the number of $H^-$ ions we now obtain

$$\log \frac{N(H)}{N(H^-)} = 0.6 - 0.48 + 10.0 - 0.35 - 3.0 \approx +6.8.$$

The ratio of $H^-/H$ is decreased by only a factor of 5 as compared to the ratio in the sun. However, if we now compute the number of hydrogen

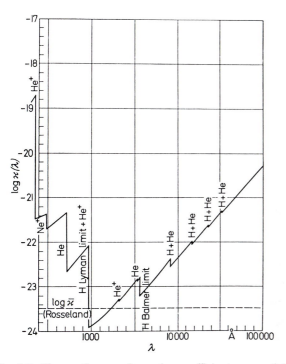

*Fig. 7.9.* The continuous absorption coefficient per particle is shown as a function of wavelength for a temperature of 28 300 K and $\log P_e = 3.5$ as appropriate for a main sequence B0 star. (From Unsöld, 1977, p. 159.)

atoms in the level with $n = 3$, i.e. in the Paschen level, we find

$$\log \frac{N_{\mathrm{H}}(n=3)}{N_{\mathrm{H}}(n=1)} = 0.95 - 12.1\Theta = 0.95 - 6.05 = -5.10;$$

this number has increased by a factor $10^4$. We now have about 100 times more hydrogen atoms in the Paschen level than we have $H^-$ ions. The absorption of neutral hydrogen is now much more important than that of $H^-$. Therefore, for A stars we find large discontinuities of $\kappa$ at the Paschen and Balmer limits.

In Fig. 7.9 we show the wavelength dependence of the absorption coefficient for a temperature of 28 300 K and an electron pressure $\log P_e = 3.5$. These values are appropriate for a B0 star on the main sequence. The hydrogen absorption coefficient is still the most important one in the visual. At very short wavelengths the helium absorption becomes noticeable. At low points in the hydrogen absorption, such as around 3700 Å, just before the Balmer absorption edge the electron scattering $\sigma_{\mathrm{el}}$ fills in. For increasing temperatures the electron scattering becomes increasingly more important.

When in the following we talk about the continuous $\kappa_c$ we always include the electron and Rayleigh scattering.

# 8

## The influence of the non-greyness of the absorption coefficient

### 8.1    The continuum energy distribution

We have seen earlier that for a linear source function

$$F_\lambda = S_\lambda(\tau_\lambda = \tfrac{2}{3}),\qquad\qquad 8.1$$

which is called the Eddington–Barbier relation. For the grey case when $\tau_\lambda = \tau$, the layer with $\tau_\lambda = \tfrac{2}{3}$ always refers to the same geometrical depth. But if $\kappa_\lambda$ is wavelength dependent this is no longer the case. Look at the $\kappa_\lambda$ in a B0 star (Fig. 7.9). At $\lambda = 3647^+$ we have a lower than average $\kappa_\lambda$; at $\lambda = 3647^-$ the $\kappa_\lambda$ is much higher than average. In Fig. 8.1 we demonstrate the situation.

At $\lambda = 3647^+$, i.e., on the long wavelength side of the discontinuity, the layer with $\tau_\lambda = \tfrac{2}{3}$ refers to a geometrical depth $t_\lambda$ which is

$$t_\lambda = \frac{2}{3}\,\frac{1}{\kappa_\lambda(3647^+)},\qquad\qquad 8.2$$

and which is larger than the average geometrical depth seen, namely

$$t = \frac{2}{3}\,\frac{1}{\bar\kappa},\qquad\qquad 8.3$$

where $\bar\kappa$ is the average absorption coefficient, averaged over wavelengths.

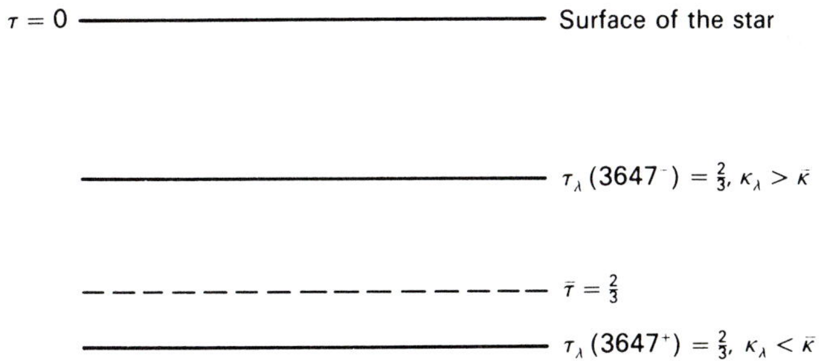

$\tau = 0$ ————————————————— Surface of the star

————————————————— $\tau_\lambda(3647^-) = \tfrac{2}{3},\ \kappa_\lambda > \bar\kappa$

– – – – – – – – – – – – – – $\bar\tau = \tfrac{2}{3}$

————————————————— $\tau_\lambda(3647^+) = \tfrac{2}{3},\ \kappa_\lambda < \bar\kappa$

*Fig. 8.1.* For wavelengths with large $\kappa_\lambda$ we see high layers with low temperatures. For wavelengths with small $\kappa_\lambda$ we see deep layers with high $T$. $\bar\tau$ is the optical depth corresponding to $\bar\kappa$.

Thus, at this wavelength of 3647$^+$ Å, we obtain radiation from a deeper than average geometrical depth. At the deeper layer the source function and the temperature are larger and so we get more radiation at this wavelength with a small $\kappa_\lambda$ than we would have received in a grey atmosphere with $\kappa = \bar{\kappa}$.

On the other hand, at $\lambda = 3647^-$ Å, i.e., at the short wavelength side of the Balmer discontinuity, where $\kappa_\lambda$ is larger than $\bar{\kappa}$, we receive radiation from a higher level, where $T$ and $S_\lambda$ are smaller than at $\bar{\tau} = \frac{2}{3}$, and we receive less radiation than for a grey atmosphere with $\kappa = \bar{\kappa}$. Here $\bar{\tau}$ is the optical depth corresponding to $\bar{\kappa}$.

We show the relation between the wavelength dependence of $\kappa_\lambda$ and $F_\lambda$ in Fig. 8.2. The distribution of the flux $F_\lambda$ looks like a mirror image of the wavelength distribution of $\kappa_\lambda$. We find discontinuities in the continuum energy distribution at those wavelengths where $\kappa_\lambda$ has a discontinuity.

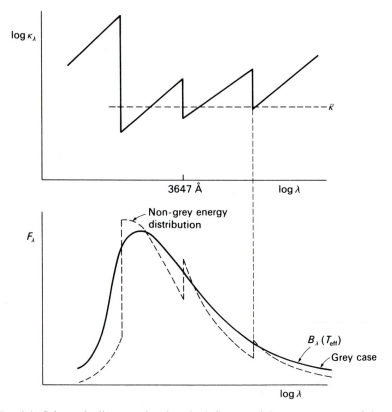

*Fig. 8.2.* Schematic diagram showing the influence of the non-greyness of the hydrogen absorption coefficient on the observed energy distribution of the star.

## 8.2 The dependence of the Balmer discontinuity on temperature and electron density

For F and G stars the discontinuity at the Balmer absorption edge is determined by the ratio of the absorption coefficient at the long wavelength side, which is due to the $H^-$ absorption, to the absorption coefficient on the short wavelength side, which is due to the sum of the $H^-$ absorption and the H absorption from the second quantum level (see Fig. 8.3). This means

$$\frac{\kappa(3647^+)}{\kappa(3647^-)} = \frac{\kappa(H^-)N(H^-)}{\kappa(H^-)N(H^-) + \kappa^-(H)N_H(n=2)}. \qquad 8.4$$

Here $\kappa^-(H)$ and $\kappa(H^-)$ are the hydrogen and $H^-$ absorption coefficients per H atom in the level $n = 2$ and per $H^-$ ion respectively at the short wavelength side of 3647 Å. From the Saha equation we know that

$$N(H^-) = f(T)N_H(n=1)n_e,$$

where $n_e$ is the number of free electrons per $cm^3$, also called the *electron density*, and $f(T)$ is a slowly varying function of $T$. We find then that

$$\frac{\kappa(3647^+)}{\kappa(3647^-)} = \frac{\kappa(H^-)N_H(n=1)n_e f(T)}{\kappa(H^-)N_H(n=1)n_e f(T) + \kappa^-(H)N_H(n=2)}. \qquad 8.5$$

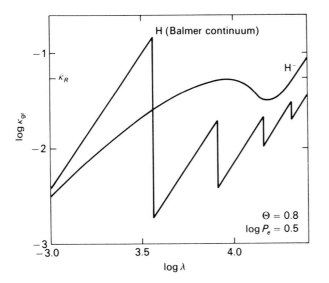

*Fig. 8.3.* The continuous absorption coefficients $\kappa$ per gram of hydrogen and the negative hydrogen ion $H^-$ for a temperature of 6300 K and an electron pressure of $\log P_e = 0.5$. For $\lambda < 3647$ Å, $\kappa$ is mainly determined by neutral hydrogen, while for $\lambda > 3647$ Å, $\kappa$ is determined by $H^-$.

Generally, the discontinuity is a measure of the temperature and the electron density, and can be used to determine one of these quantities.

For F stars the contribution of H$^-$ to the absorption on the short wavelength side of $\lambda = 3647$ Å is very small. In this case, equation (8.5) reduces to

$$\frac{\kappa(3647^+)}{\kappa(3647^-)} = \frac{\kappa(H^-)N_H(n=1)n_e f(T)}{\kappa^-(H)N_H(n=2)} \propto \frac{N_H(n=1)}{N_H(n=2)} n_e. \qquad 8.6$$

*For cool stars the Balmer discontinuity depends on both the temperature and the electron density.* If we can determine the temperature from another observation we can determine the electron density from this discontinuity. With increasing electron density the discontinuity becomes smaller because the H$^-$ absorption increases when more H$^-$ ions are formed for higher electron densities. In Fig. 8.4 we compare the energy distributions of an

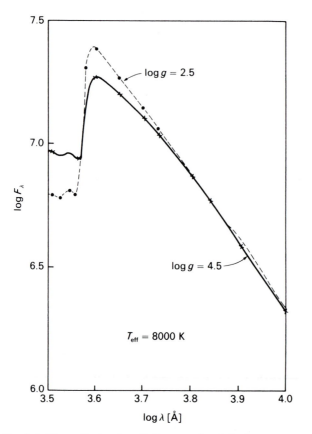

*Fig. 8.4.* The radiative flux $F_\lambda$ [erg cm$^{-2}$ s$^{-1}$ Å$^{-1}$] is shown as a function of wavelength for a main sequence star with $\log g = 4.5$ (solid line) and for a bright giant with $\log g = 2.5$ (dashed curve). The bright giant shows a much larger Balmer jump because of the smaller contribution of H$^-$ to the absorption coefficient.

F0 V star with one for an F0 II bright giant. The Balmer discontinuity is larger for the bright giant.

For stars hotter than 9000 K the $H^-$ absorption is negligible in comparison with the hydrogen absorption. The absorption coefficient on the long wavelength side of the Balmer discontinuity, also called the *Balmer jump*, is due to the absorption from hydrogen atoms in the third quantum level, while that on the short wavelength side is mainly due to hydrogen absorption from atoms in the second quantum level. The ratio of the absorption coefficients then becomes

$$\frac{\kappa(3647^+)}{\kappa(3647^-)} \approx \frac{\kappa^+(H)N_H(n=3)}{\kappa^-(H)N_H(n=2)}, \qquad 8.7$$

where $\kappa^+(H)$ is the hydrogen absorption coefficient per absorbing atom at the long wavelength side of the Balmer jump and $\kappa^-(H)$ that on the short wavelength side. For the higher temperatures the Balmer jump depends only on the ratio of the hydrogen atoms in the second and in the third quantum levels, which, according to the Boltzmann formula, is only a function of temperature. *For hot stars the temperature can therefore be determined from the Balmer jump alone.*

For yet higher temperatures and for supergiants the electron scattering becomes important first on the long wavelength side and for still higher temperatures also on the short wavelength side of the Balmer discontinuity. This reduces the Balmer discontinuity. In this case the discontinuity depends again on both the temperature and the electron density.

## 8.3    The influence of the Balmer jump on the UBV colors

Fig. 1.7 shows the two-color diagram for black bodies of different temperatures. For a grey atmosphere for which we obtain radiation from the same depth $t(\tau = \frac{2}{3})$ at all wavelengths we would see radiation from the temperature at this depth at all wavelengths. If $S_\lambda = B_\lambda(T)$, then the observed energy distribution should be that of a black body with the temperature of that layer. As we saw in Section 6.4, the temperature in $\tau = \frac{2}{3}$ is equal to the effective temperature. For a grey atmosphere the energy distribution should therefore correspond to that of a black body with $T = T_{\mathrm{eff}}$. A look at Fig. 1.7 shows that the colors of the stars do not correspond to those of black bodies. Obviously, the stars do not have grey atmospheres. We have just seen how large the variations of the continuous absorption coefficients can be. In Fig. 8.2 we see how the actual stellar energy distributions deviate from those of black bodies. We receive more radiation for the wavelength regions where $\kappa_\lambda$ is small, which is generally the case on the long wavelength side of the Balmer discontinuity. In the blue region therefore, we generally receive more energy than we would in the case of a grey atmosphere. This is not the case in the visual region. We therefore expect non-grey atmospheres to look more blue than the grey ones, at least if we consider only the energy distribution in the continuum.

For wavelengths shorter than 3647 Å the energy is reduced by large factors when we look at the spectra of B, A, and F stars. This is due to the strong absorption from the hydrogen atoms in the second quantum level for these stars. The U filter, used to measure the U colors, is sensitive for radiation on both sides of the Balmer discontinuity. While there is somewhat more radiation on the long wavelength side of the Balmer discontinuity, the flux on the short wavelength side is greatly reduced. This reduction by far outweighs the increase on the long wavelength side. Because of the Balmer discontinuity the U − B colors become redder, i.e., more positive, while the B − V colors become smaller.

The changes of the colors are largest for A stars with the strong Balmer jump. In Fig. 8.5 we have qualitatively indicated by solid arrows which way the continuum colors change due to the Balmer absorption.

Additional strong variations of $\kappa_\lambda$ with wavelength are introduced by the spectral lines in which the absorption coefficient also changes by very large factors. While we will discuss the formation of spectral lines in Chapters 10 and 11, we discuss here which color changes are introduced due to the spectral lines. We have already seen that they are strongest for the reddest or coolest stars. The colors of these stars are therefore strongly influenced by the spectral line absorption. We also observe that there are more lines in the blue region than in the visual and that there are still more lines in the ultraviolet. The line absorption therefore tends to make all colors more red. This color change is superimposed on the color change due to the variation in the continuous $\kappa$. It becomes very strong for stars with spectral types later than F. It is of little influence for stars with spectral types earlier than F. In Fig. 8.5 we have indicated by dashed arrows the color changes for cool stars due to line absorption.

## 8.4    The influence of the non-greyness on the temperature stratification

### 8.4.1    *The mean absorption coefficient $\bar{\kappa}$*

We may expect that the grey temperature stratification is still a good approximation if we choose an appropriate mean value for the $\bar{\kappa}$ and the corresponding $\bar{\tau}$. How do we have to determine $\bar{\kappa}$ in order to make the grey temperature stratification a good approximation? Since the flux determines the temperature stratification, the flux ought to come out correctly. So we must require that $\int_0^\infty F_\lambda \, d\lambda$ is the same for the non-grey case and in the grey approximation.

As in the grey case, we want to write

$$S = B = \tfrac{3}{4}F(\bar\tau + q(\bar\tau)) \simeq \tfrac{3}{4}F(\bar\tau + \tfrac{2}{3}) \qquad \text{or} \qquad F = B(\bar\tau = \tfrac{2}{3}). \qquad 8.8$$

For a linear source function, i.e., $S_\lambda = a + b\tau_\lambda$, we had seen earlier that at the surface

$$F = \int_0^\infty F_\lambda \, d\lambda = \int_0^\infty B_\lambda(\tau_\lambda = \tfrac{2}{3}) \, d\lambda. \qquad 8.9$$

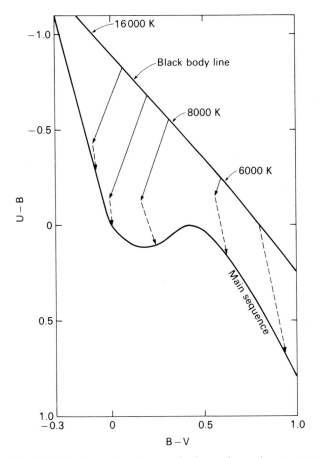

Fig. 8.5. The two-color diagram is shown for main sequence stars and for black bodies. The solid arrows show qualitatively how the colors of grey stars (black bodies) change due to the Balmer discontinuity. The dashed arrows indicate how the line absorption changes the colors qualitatively. For hot stars the Balmer discontinuity is more important than the line absorption. For cool stars the line absorption becomes very important, while the Balmer discontinuity disappears.

This should then be equal to what is obtained in the grey case:

$$\int_0^\infty F_\lambda \, d\lambda = \int_0^\infty B_\lambda(\bar\tau = \tfrac{2}{3}) \, d\lambda. \tag{8.10}$$

As can be seen from Fig. 8.6, we can describe $B_\lambda(\tau_\lambda = \tfrac{2}{3})$ by a Taylor expansion around the layer $\bar\tau = \tfrac{2}{3}$. One then obtains

$$B_\lambda(\tau_\lambda = \tfrac{2}{3}) = B_\lambda(\bar\tau = \tfrac{2}{3}) + \frac{dB_\lambda}{d\bar\tau} \Delta\bar\tau, \tag{8.11}$$

where the derivative has to be taken at the depth $\bar\tau = \tfrac{2}{3}$. Assuming $\kappa_\lambda/\bar\kappa$ to be the depth independent, we derive

$$\frac{\tau_\lambda}{\bar\tau} = \frac{\kappa_\lambda}{\bar\kappa} \qquad \text{and} \qquad \bar\tau = \tau_\lambda \frac{\bar\kappa}{\kappa_\lambda}. \tag{8.12}$$

We then find that at $\tau_\lambda = \tfrac{2}{3}$

$$\bar\tau = \frac{2}{3}\frac{\bar\kappa}{\kappa_\lambda} = \frac{2}{3} + \Delta\bar\tau, \tag{8.13}$$

according to the definition of $\Delta\bar\tau$ (see Fig. 8.6). From this equation we derive

$$\Delta\bar\tau = \frac{2}{3}\frac{\bar\kappa}{\kappa_\lambda} - \frac{2}{3} = \frac{2}{3}\left(\frac{\bar\kappa}{\kappa_\lambda} - 1\right). \tag{8.14}$$

The condition (8.9) for $F_\lambda$ then requires

$$\int_0^\infty B_\lambda(\tau_\lambda = \tfrac{2}{3}) \, d\lambda = \int_0^\infty B_\lambda(\bar\tau = \tfrac{2}{3}) \, d\lambda + \int_0^\infty \frac{dB_\lambda}{d\bar\tau} \times \frac{2}{3}\left(\frac{\bar\kappa}{\kappa_\lambda} - 1\right) d\lambda$$

$$= \int_0^\infty B_\lambda(\bar\tau = \tfrac{2}{3}) \, d\lambda. \tag{8.15}$$

For this to hold, we must require

$$\int_0^\infty \frac{dB_\lambda}{d\bar\tau} \frac{2}{3}\left(\frac{\bar\kappa}{\kappa_\lambda} - 1\right) d\lambda = 0 \qquad \text{or} \qquad \int_0^\infty \frac{dB_\lambda}{d\bar\tau} \frac{\bar\kappa}{\kappa_\lambda} \, d\lambda = \int_0^\infty \frac{dB_\lambda}{d\bar\tau} \, d\lambda, \tag{8.16}$$

Fig. 8.6 demonstrates that $B_\lambda(\tau_\lambda = \tfrac{2}{3}) = B_\lambda(\bar\tau = \tfrac{2}{3}) + \dfrac{dB_\lambda}{d\tau_\lambda} \Delta\bar\tau(\tau_\lambda = \tfrac{2}{3})$, with $\Delta\bar\tau < 0$.

or, after dividing by $\bar{\kappa}$,

$$\frac{1}{\bar{\kappa}} \int_0^\infty \frac{dB_\lambda}{d\bar{\tau}} \, d\lambda = \int_0^\infty \frac{1}{\kappa_\lambda} \frac{dB_\lambda}{d\bar{\tau}} \, d\lambda \qquad \text{with} \quad \frac{dB_\lambda}{d\bar{\tau}} = \frac{dB_\lambda}{dT} \frac{dT}{d\bar{\tau}} . \qquad 8.17$$

This is the prescription for the determination of $\bar{\kappa}$, which we have derived here for the surface, for a linear depth dependence of the source function, and depth independent $\kappa_\lambda / \bar{\kappa}$.

The same condition can be derived for deeper layers if we start from the transfer equation (5.3) after integration over $d\omega$:

$$\frac{d}{d\tau_\lambda} \tfrac{1}{4} F_\lambda = J_\lambda - S_\lambda; \qquad 8.18$$

or, if we multiply first by $\cos \vartheta$ and then integrate over $\omega$ and divide by $\pi$ we find with the Eddington approximation $K_\lambda = \tfrac{1}{3} J_\lambda$ (which holds for an isotropic radiation field) that

$$\frac{dK_\lambda}{d\tau_\lambda} = \tfrac{1}{4} F_\lambda = \frac{1}{3} \frac{dB_\lambda}{d\tau_\lambda} , \qquad 8.19$$

for a deeper layer, where $J_\lambda \approx B_\lambda$. Or, rewriting the second half of the equation, we find

$$F_\lambda = \frac{4}{3} \frac{dB_\lambda}{d\tau_\lambda} . \qquad 8.20$$

This is a very important relation between $F_\lambda$ and the gradient of the source function.

From the condition that

$$\int_0^\infty F_\lambda \, d\lambda = F \text{ (grey)}, \qquad 8.21$$

we now derive

$$\frac{3}{4} \int_0^\infty F_\lambda \, d\lambda = \int_0^\infty \frac{dB_\lambda}{d\tau_\lambda} \, d\lambda = \int_0^\infty \frac{dB_\lambda}{d\bar{\tau}} \, d\lambda \qquad \text{with} \quad \tfrac{3}{4} F = \frac{dB}{d\bar{\tau}} . \qquad 8.22$$

This yields

$$\int_0^\infty \frac{1}{\kappa_\lambda} \frac{dB_\lambda}{dt} \, d\lambda = \frac{1}{\bar{\kappa}} \int_0^\infty \frac{dB_\lambda}{dt} \, d\lambda, \qquad 8.23$$

which is the same prescription as above. Using again

$$\frac{dB_\lambda}{dt} = \frac{dB_\lambda}{dT} \frac{dT}{dt} \qquad \text{and} \qquad \frac{dB}{dt} = \frac{dB}{dT} \frac{dT}{dt} ,$$

we derive again

$$\frac{1}{\bar{\kappa}} = \frac{\displaystyle\int_0^\infty \frac{1}{\kappa_\lambda} \frac{\mathrm{d}B_\lambda}{\mathrm{d}T} \, \mathrm{d}\lambda}{\displaystyle\int_0^\infty \frac{\mathrm{d}B_\lambda}{\mathrm{d}T} \, \mathrm{d}\lambda} = \int_0^\infty \frac{1}{\kappa_\lambda} G'_\lambda(T) \, \mathrm{d}\lambda = \int_0^\infty \frac{1}{\kappa_\lambda} G(\alpha) \, \mathrm{d}\alpha, \qquad 8.24$$

where $G'_\lambda(T)$ is the Rosseland weighting function and $\alpha = h\nu/kT$. From this definition of $G(\alpha)$ we find

$$G(\alpha) = \frac{kT}{h} \frac{\lambda^2}{c} \frac{\mathrm{d}B_\lambda/\mathrm{d}T}{\mathrm{d}B/\mathrm{d}T} = \frac{15}{4\pi^4} \frac{\alpha^4 \mathrm{e}^\alpha}{(\mathrm{e}^\alpha - 1)^2}. \qquad 8.25$$

This kind of mean absorption coefficient is called the *Rosseland mean absorption coefficient*. It is a harmonic mean. Frequencies with small $\kappa_\lambda$ are weighted most heavily, because the *Rosseland mean $\bar{\kappa}$ is a flux weighted mean* and the flux is largest in the regions of small $\kappa_\lambda$, as we see from equation (8.20) if we write it in the form

$$F_\lambda = \frac{1}{\kappa_\lambda} \frac{\mathrm{d}B_\lambda}{\mathrm{d}t} \frac{4}{3}. \qquad 8.26$$

Because of this, the Rosseland mean $\bar{\kappa}$ is always close to the smallest values of $\kappa_\lambda$. See Figs. 7.5 and 8.3, where we have indicated the Rosseland mean values $\bar{\kappa}_R$.

With the prescription (8.24) we can calculate mean values of $\bar{\kappa}$ for all desired temperatures and electron pressures.

For a given temperature and electron pressure we can calculate the degree of ionization for all the atoms and ions present in stellar atmospheres. With this we can add up all the numbers of particles to obtain the total number of particles per $cm^3$ from which the gas pressure can be obtained. We can thus also determine $\bar{\kappa}$ as a function of temperature and gas pressure. In Fig. 8.7 we reproduce a plot showing $\bar{\kappa}$ as a function of the gas pressure for different values of the temperature. The values of $\Theta = 5040/T$ are given on each curve. For low temperatures the continuous $\kappa$ is essentially due to $H^-$ only. For a given temperature the average $\bar{\kappa}$ increases proportionally to the electron pressure. For very low pressures the electron scattering becomes important and prevents a further decrease of $\bar{\kappa}$. At temperatures above 6000 K hydrogen absorption becomes important for very low gas pressures. At temperatures above 10 000 K hydrogen absorption is important for all pressures considered here and gives a large increase in $\bar{\kappa}$. At still higher temperatures hydrogen ionizes and $\bar{\kappa}$ does not increase any more, but decreases with increasing $T$. Finally, electron scattering takes over.

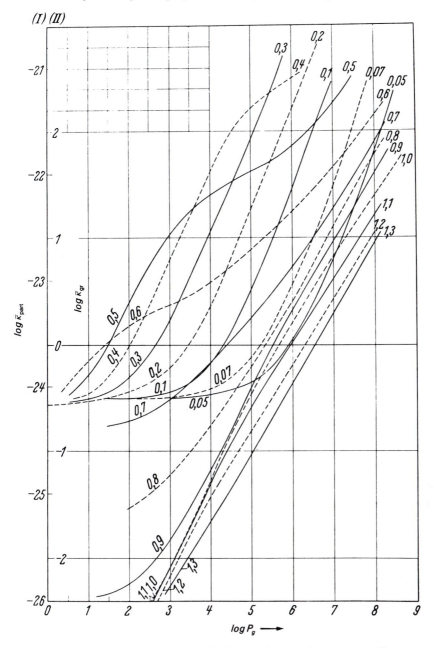

*Fig. 8.7.* The Rosseland mean values $\bar{\kappa}_R$ for the continuous absorption coefficient are shown as a function of the gas pressure. Each curve corresponds to a given value of the temperature. The value of $\Theta = 5040/T$ is given for each curve. The plot is double logarithmic. What is shown is actually the absorption coefficient per gram of material, i.e., for a column of gas which contains one gram of material, or per heavy particle, as shown on the outer left-hand scale. Solar element abundances were assumed.

### 8.4.2    Changes in the temperature stratification $T(\bar{\tau})$ due to the non-greyness

By computing $\bar{\kappa}$ in such a way as to make the integrated flux the same for the grey and for the non-grey case, we hoped that the temperature stratification

$$T^4 = \tfrac{3}{4}T_{\text{eff}}^4(\bar{\tau}_R + q(\bar{\tau}_R)),\qquad\qquad 8.27$$

with $\bar{\tau}_R$ being the Rosseland mean value of $\tau_\lambda$, i.e., with

$$\bar{\tau}_R = \int_0^t \bar{\kappa}_R \, dt,\qquad\qquad 8.28$$

will still give a reasonable approximation. This is true qualitatively. Some important modifications, however, have to be taken into account.

We can see what happens if we look at the condition for radiative equilibrium

$$\int_0^\infty \kappa_\lambda J_\lambda \, d\lambda = \int_0^\infty \kappa_\lambda S_\lambda \, d\lambda \rightarrow = \int_0^\infty \kappa_\lambda B_\lambda \, d\lambda \qquad \text{for LTE.} \qquad 8.29$$

If $\kappa_\lambda$ varies strongly with $\lambda$, then these integrals are mainly determined by the wavelengths with *large* $\kappa_\lambda$. What happens to $J_\lambda$ when $\kappa_\lambda$ becomes very large as compared to the grey case? Let us look at the surface first, where $I_\lambda(\cos\vartheta) \sim B_\lambda(\tau_\lambda = \cos\vartheta)$. With the large $\kappa_\lambda$, the $I_\lambda$ now comes from higher layers and therefore is smaller. So $J_\lambda$ will also become smaller, but we must require $\int_0^\infty \kappa_\lambda J_\lambda \, d\lambda = \int_0^\infty \kappa_\lambda B_\lambda \, d\lambda$. If $J_\lambda$ becomes smaller, $B_\lambda$ must also become smaller at the surface, which means the temperature must decrease at the surface.

Therefore, generally the first effect of the non-greyness is a decrease in surface temperature; this is also called *surface cooling* and is due to the larger emission because of the larger $\kappa_\lambda$ (though under special circumstances this may not be the case as for instance for strong lines in the red).

The second effect is the *backwarming*. We now have essentially many strong lines in the spectrum. In the regions with large $\kappa_\lambda$ the intensity $I_\lambda$ and the flux $F_\lambda$ decrease, which means the total flux coming out of an atmosphere with the same temperature stratification is decreased. This means the same temperature stratification now leads to a smaller effective temperature. If we want to obtain the same total flux, we have to get more flux out of the wavelengths with the small $\kappa_\lambda$. If we have mainly many lines the flux has to be obtained from the wavelengths between the lines. This can only be done by increasing the temperature in the atmosphere.

Some people call this redistribution of energy: the energy absorbed at some wavelengths comes out at other wavelengths. But remember, this can only be done by way of increasing the temperature in deeper layers. The energy distribution in the continuum now resembles the energy distribution of a somewhat hotter star, if the non-greyness is due mainly to line absorption. The star cannot absorb the energy at one wavelength and arbitrarily emit it at another wavelength, as we please.

A modified temperature distribution of the form

$$T^4 = \tfrac{3}{4}T'^4_{\text{eff}}(\bar{\tau} + q(\bar{\tau}))$$  8.30

leads to a flux

$$\int_0^\infty \pi F_\lambda \, d\lambda = \pi F = \sigma T'^4_{\text{eff}}(1 - \eta),$$  8.31

if $\eta$ is the fraction of energy absorbed by the lines. If we require $\pi F = \sigma T^4_{\text{eff}}$ as before, we must have

$$T^4_{\text{eff}} = T'^4_{\text{eff}}(1 - \eta) \qquad \text{or} \qquad T'^4_{\text{eff}} = \frac{T^4_{\text{eff}}}{1 - \eta}.$$  8.32

In the case of the sun, $\eta \sim 14\%$, so $(1 - \eta) = 0.86$ and

$$T'_{\text{eff}} = T_{\text{eff}}/\sqrt[4]{0.86} \approx T_{\text{eff}}(1 + 0.037).$$  8.33

In the sun with $\eta \sim 0.14$ the $T'_{\text{eff}}$ must be higher than $T_{\text{eff}}$ by about 3.5%, or by about 200 degrees. In the continuum the atmosphere looks like one

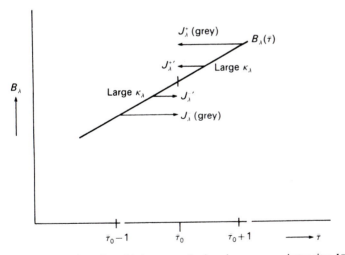

Fig. 8.8. In wavelengths with large $\kappa_\lambda$ the ingoing average intensity $J_\lambda^-$ is increased by $\Delta J^-$ because the radiation originates at a deeper point with a larger Planck function than in the grey case. The outgoing radiation originates at a higher layer with a smaller Planck function than in the grey case. The average intensity $J_\lambda^+$ is reduced by the amount $\Delta J^+$ as compared to the grey case. If the Planck function is a linear function of depth $\Delta J_\lambda^+ = -\Delta J_\lambda^-$ and for deep layers the sum $J_\lambda^+ + J_\lambda^-$ remains the same as in the grey case.

with a temperature increased by about 200 degrees. With an effective temperature of 5800 K for the sun the energy distribution in the continuum looks similar to that of a black body of 6000 K, as we saw earlier in Volume 1.

What happens to the $J_\lambda$ in the deep layers?

For wavelengths with small $\kappa_\lambda$ the outgoing average intensity $J_\lambda^+$ is increased because it now originates in deeper layers than in the grey case, but the inward going intensity now originates at a higher layer with a smaller Planck function. The sum of the ingoing and outgoing intensities remains the same as long as we are far enough away from the surface such that the Planck function is still a linear function of depth. See Fig. 8.8 for the case of large $\kappa_\lambda$.

# 9

# The pressure stratification

## 9.1    The hydrostatic equilibrium equation

For the computation of the $\kappa_\lambda$ we have to know the number of absorbing atoms per $cm^3$ of hydrogen or of the metals. This number is determined from the Boltzmann and Saha equations, but the Saha equation requires the knowledge of the electron pressure and also the number densities of the heavy particles if we want to know how many ions or atoms of a given kind we have per unit volume. In order to compute these number densities we have to know the gas pressure. How can that be determined?

The first question is, why doesn't our atmosphere escape into space? Obviously, the atmosphere is unbound with a vacuum outside. Gravity must prevent it from escaping. The next question is, why doesn't gravity pull the atmosphere down to the surface of the Earth? Because the gas pressure prevents it. Obviously, the atmosphere can only be stable if gravitational and pressure forces balance, i.e. if it is in *hydrostatic equilibrium*.

Let us look at the pressure force on a given volume of gas (see Fig. 9.1a). Let $t$ be the coordinate of depth in the atmosphere. At the depth $t$,

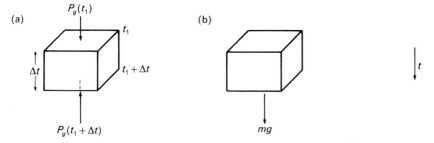

Fig. 9.1. (a) The net pressure force on a given volume of gas is the difference between the pressure force pushing up from the bottom and the one pushing down from the top. (b) The gravitational force equals the weight of the volume element under consideration. In an atmosphere pressure forces and gravitational forces must balance.

101

we consider a volume element of vertical extent $\Delta t$. From the top the gas pressure $P_g(t_1)$ presses the element down; from the bottom the gas pressure $P_g(t_1 + \Delta t)$ tries to push up the volume of gas considered. The net pressure force on the volume element under consideration is then

$$\Delta P_g = P_g(t_1 + \Delta t) - P_g(t_1) = P_g(t_1) + \frac{dP_g}{dt}\Delta t - P_g(t_1) = \frac{dP_g}{dt}\Delta t. \qquad 9.1$$

If the cross-section of the column is $1\,\mathrm{cm}^2$, the gravitational force is

$$gm = g\rho\Delta V = g\rho\Delta t \qquad \text{since } \Delta V = 1\,\mathrm{cm}^2 \times \Delta t, \qquad 9.2$$

where $m$ is the mass of the volume element, $\rho$ is the density, $g$ is the gravitational acceleration, and $\Delta V$ is the volume under consideration. The balance of gravitational and pressure force then requires

$$\frac{dP_g}{dt}\Delta t = g\rho\Delta t \qquad \text{or} \qquad \frac{dP_g}{dt} = g\rho. \qquad 9.3$$

This is called the *hydrostatic equation*.

Replacing $\rho$ by $P_g\mu m_H/(kT)$, we obtain with $k/m_H = R_g$, $m_H = $ mass of the hydrogen atom, $R_g = $ gas constant,

$$\frac{dP_g}{dt} = \frac{gP_g\mu}{R_g T} \qquad \text{or} \qquad \frac{d\ln P_g}{dt} = \frac{g\mu}{R_g T}. \qquad 9.4$$

For the calculation of $P_g(t)$ we generally need to know the temperature $T$ as a function of $t$. For an isothermal atmosphere with $T = $ const. and also $\mu = $ const., we can, however, integrate this equation and obtain

$$\ln P_g - \ln P_{g0} = \frac{g\mu}{R_g T}(t - t_0) = \frac{g\mu}{R_g T}\Delta t, \qquad 9.5$$

or, taking the exponential of this equation, we find

$$P_g = P_{g0}\,e^{[(g\mu/R_g T)\Delta t]} = P_{g0}\,e^{\Delta t/H} \qquad \text{with } H = \frac{R_g T}{\mu g}. \qquad 9.6$$

$H$ is called the *scale height*. In an isothermal atmosphere, the pressure changes by a factor of e over 1 scale height. We also compute that for a fictitious atmosphere of constant density $\rho_0$, corresponding to the gas pressure $P_{g0}$ at the base of the real atmosphere, we can put the total mass of the real atmosphere into a layer of height $H$. Therefore $H$ is also called the *equivalent height* of the atmosphere.

When using the Saha equation, as we need to do for the calculation of $P_e$ and $\kappa_\lambda$, we need temperature and pressure in a given layer of the atmosphere. We can describe the position of this layer by its geometrical depth $t$ or by its optical depth $\tau_\lambda$ or $\bar{\tau}$. Since we know the temperature as a function of $\bar{\tau}$, it is convenient to compute the gas pressure also as a function of $\bar{\tau}$. In order to do this we divide equation (9.3) by $\bar{\kappa}_{cm}$ and obtain with $d\bar{\tau} = \bar{\kappa}_{cm}\, dt$

$$\frac{dP_g}{d\bar{\tau}} = \frac{g\rho}{\bar{\kappa}_{cm}} = \frac{g}{\bar{\kappa}_{cm}/\rho}. \qquad 9.7$$

$\bar{\kappa}_{cm}$ stands for $\bar{\kappa}$ per cm$^3$, which has the dimension cm$^{-1}$.

Here we need to pause and consider the units of the absorption coefficient $\kappa$. We introduced $\kappa$ when we considered the change of the intensity $I$ along a path of length $\Delta s$; we wrote $\Delta I/\Delta s = -\kappa I$. The dimension of $\kappa$ must therefore be cm$^{-1}$. It describes the fractional change of intensity per cm. We note that the dimension and the value of $\kappa$ does not depend on whether we use the intensity $I_\nu$ or $I_\lambda$ because $\kappa$ gives the *fractional* change. $\kappa_{cm} = \kappa$ [cm$^{-1}$]* obviously depends on the number of particles in 1 cm$^3$, which changes with increasing density. It is often much better to use a quantity which is less density dependent. We therefore often describe the change of intensity which has occurred after the beam of light has passed through a column of gas with a cross-section of 1 cm$^2$ and which contains 1 gram of material. If no change in the degree of ionization occurs, then the number of particles in 1 gram of material is always the same for a given chemical composition. The $\kappa$ for this amount of material therefore depends much less on temperature and pressure than does the $\kappa_{cm}$. How long is the column of gas which contains 1 gram of material? If the density is $\rho$ then the length $s$ of the column has to be $s = 1/\rho$. Clearly, $\kappa$ per gram $= \kappa_{gr} = \kappa_{cm}\, s$, where $s$ is the length of the column containing 1 gram of material. The fractional change along this column is therefore $\kappa_{cm}\, s$ or $\kappa$ per gram $= \kappa$ per cm$^3$ divided by $\rho$. We write this as

$$\kappa_{gr} = \kappa_{cm}/\rho. \qquad 9.8$$

$\kappa_{gr}$ has the dimension cm$^2$ g$^{-1}$, which is that of a cross-section. Many astronomers, especially those that discuss stellar interiors, mean $\bar{\kappa}_{gr}$ if they just write $\bar{\kappa}$.

Now we can go back to equation (9.7). On the right-hand side we have in the denominator the ratio $\kappa_{cm}/\rho$, which can be replaced by $\kappa_{gr}$ and obtain

$$\frac{dP_g}{d\bar{\tau}} = \frac{g}{\bar{\kappa}_{gr}}. \qquad 9.9$$

The gas pressure $P_g(\bar{\tau})$ can be obtained by integrating this differential equation.

## 9.2 Integration of the hydrostatic equilibrium equation

In a formal way we can integrate the hydrostatic equilibrium equation by writing

---

* $\kappa_{cm} = \kappa$ per particle [cm$^2$] $\times N$ [cm$^{-3}$] $= \kappa$ per cm$^3$, where $N$ = number of particles per cm$^3$.

$$P_g - P_{g0} = \int_{\bar{\tau}_0}^{\bar{\tau}} \frac{g}{\bar{\kappa}_{gr}} \, d\bar{\tau}, \qquad P_{g0} = P_g(\bar{\tau}_0). \qquad\qquad 9.10$$

Here $\bar{\kappa}_{gr}$ is a complicated function of temperature and pressure (see Fig. 8.7). We would have to know the pressure in order to calculate the right-hand side. In general, the integral can be evaluated by stepwise integration, as we will soon discuss, but let us first use some simple approximations for $\bar{\kappa}_{gr}$ as a function of optical depth or of pressure. These permit an analytical integration of the differential equation (9.9). The simplest approximation is the assumption $\bar{\kappa}_{gr} = $ const. This is not as unreasonable as it may seem at first sight, since the number $N$ of particles per gram of material is constant for a given chemical composition provided the degree of ionization does not change. It is

$$N = \frac{1}{\mu m_{\mathrm{H}}}.$$

Of course, as we saw earlier, even for the same number of particles $\bar{\kappa}$ may vary greatly by changing degrees of excitation. So $\bar{\kappa}_{gr} = $ const. is not a good approximation, but let us see what we obtain.

Integration of equation (9.9) yields

$$P_g - P_{g0} = \frac{g}{\bar{\kappa}_{gr}} (\bar{\tau} - \bar{\tau}_0). \qquad\qquad 9.11$$

With $\tau_0 = 0$ and $P_{g0} = 0$ we find

$$P_g = \frac{g}{\bar{\kappa}_{gr}} \bar{\tau}. \qquad\qquad 9.12$$

Knowing $T(\bar{\tau})$ for a given $T_{\mathrm{eff}}$, we can assume a value for $\bar{\kappa}_{gr}$, insert this into equation (9.12), and compute a value for the gas pressure. With this first approximation to the gas pressure a value of $\bar{\kappa}_{gr}$ can now be calculated. Inserting this into equation (9.12) gives a better approximation to the gas pressure. With this we can calculate a better value for $\bar{\kappa}_{gr}$ and so on until we achieve convergence, which means until $\bar{\kappa}_{gr}$ and $P_g$ are consistent. Of course, this does not mean that we have the correct gas pressure at the depth $\bar{\tau}$, since equation (9.12) is based on the erroneous assumption that $\bar{\kappa}_{gr}$ is constant with depth. Nevertheless, this method is used in all modern computer programs to find a starting value $P_{g0}$ in equation (9.10). If we were to follow the iteration procedure as described, it would take rather long before convergence is achieved. There are various ways to speed up the convergence, but the basic method remains the same.

Because the value $P_{g0}$ obtained in this way is not the correct one, we have to make sure that $P_{g0}$ refers to a very high layer in the atmosphere such that $P_{g0} \ll P_g(\tau)$ for any of the layers which are of interest to us. Numerical integrations are usually started at an optical depth $\bar{\tau} = 10^{-4}$ or $\bar{\tau} = 10^{-5}$.

Once we have a starting value $P_{g0}(\bar{\tau}_0)$ for the integration we can continue stepwise. Knowing $T_0(\bar{\tau}_0)$ and $P_{g0}$, we can, for this depth, calculate $\bar{\kappa}_{gr,0}(T_0, P_{g0})$. Assuming that $\bar{\kappa}_{gr}$ remains constant over the next small interval $\Delta\bar{\tau}$, we find for $\bar{\tau}_1 = \bar{\tau}_0 + \Delta\bar{\tau}$

$$P_g(\bar{\tau}_1) = P_{g0} + \frac{g}{\bar{\kappa}_{gr,0}} (\bar{\tau}_1 - \bar{\tau}_0).\qquad 9.13$$

At $\bar{\tau}_1$ we know $T(\bar{\tau}_1)$ and now also $P_{g_1} = P_g(\bar{\tau}_1)$. We can therefore determine $\bar{\kappa}_{gr,1} = \bar{\kappa}_{gr}(\bar{\tau}_1)$. In order to improve the value for $P_g(\bar{\tau}_1)$ we can now use the average value of $\bar{\kappa}_{gr}$ in the interval between $\bar{\tau}_0$ and $\bar{\tau}_1$ and insert that into equation (9.13) to obtain a better value for $P_{g_1}$. There are more sophisticated methods to get the best possible numerical value for $P_{g_1}$. Knowing $P_{g_1}$ and $\bar{\kappa}_{gr,1}$ we can now proceed in the same way to obtain $P_g(\bar{\tau}_2) = P_{g_2}$ and then $\bar{\kappa}_{gr,2} = \kappa_{gr}(\bar{\tau}_2)$, where $\bar{\tau}_2 = \bar{\tau}_1 + \Delta\bar{\tau}$. This value can then be improved in the same way $P_{g_1}$ was. We can then proceed to the next $\bar{\tau}$ and so on. In this way we can obtain very good numerical solutions of equation (9.9). There are many methods to give more accurate solutions than the one described, but the principle is the same. However, these numerical solutions do have a disadvantage in that it is difficult to estimate how the solutions are going to differ for different input parameters, such as the gravity. This can be seen much better from analytical solutions.

## 9.3 The dependence of the gas pressure on the gravitational acceleration

We said previously that the approximation $\bar{\kappa}_{gr} = $ const. is not a good one. There are other better approximations to the depth dependence of $\bar{\kappa}_{gr}$, which we shall discuss now. Generally, $\bar{\kappa}_{gr} = \bar{\kappa}_{gr}(T, P_g)$. In a given atmosphere the temperature $T$ is a given function of the optical depth $\bar{\tau}$, and so is the gas pressure $P_g$ (see Fig. 9.2). For each value of $\bar{\tau}$ both $T$ and $P_g$ are given. We can now introduce $P_g$ as the independent variable (see Fig. 9.3). For each value of $\bar{\tau}$ we determine $T$ and $P_g$ and plot in Fig. 9.3 the temperature as a function of $P_g$. For each set $T(\bar{\tau})$ and $P_g(\bar{\tau})$ we can also calculate $\bar{\kappa}(\bar{\tau})$ and also plot this as a function of $P_g(\bar{\tau})$. This $\bar{\kappa}$ appears then as a function of $P_g$ alone. Nevertheless, it is the correct $\bar{\kappa}_{gr}$ because it was calculated using the known temperature at this optical depth and pressure. Looking now at Fig. 9.3, we can approximate the *correct* $\bar{\kappa}_{gr}(P_g)$

by a simple function; for instance, by a polynomial such as

$$\bar{\kappa}_{gr} = \kappa_{gr,0} P_g^n, \qquad\qquad 9.14$$

where $n$ is an integer.

$n = 1$ usually gives a very good approximation. Inserting (9.14) into the hydrostatic equation gives

$$\frac{dP_g}{d\bar{\tau}} = \frac{g}{\kappa_{gr,0} P_g^n} \qquad \text{or} \qquad P_g^n \, dP_g = \frac{g}{\kappa_{gr,0}} \, d\bar{\tau}. \qquad 9.15$$

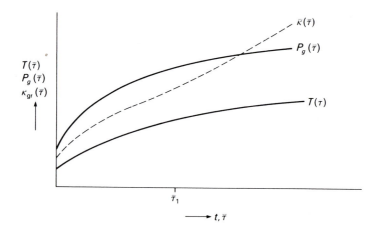

*Fig. 9.2.* In a given atmosphere temperature and pressure are given as a function of optical or of geometrical depth. For each depth we can then calculate $\bar{\kappa}_{gr}(T, P_g)$ and plot $\bar{\kappa}_{gr}$ also as a function of geometrical or optical depth.

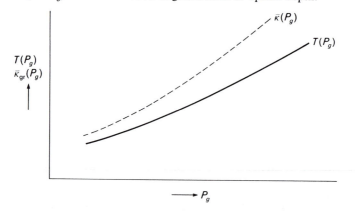

*Fig. 9.3.* For a given value of $\bar{\tau}_1$ in Fig. 9.2 we read off $T_1(\bar{\tau}_1)$ and $P_{g_1}(\bar{\tau}_1)$, and plot $T_1(P_{g_1})$. For this same value of $P_{g_1}$ we determine $\bar{\kappa}_{gr}(\bar{\tau}_1)$ from Fig. 9.2 and plot $\bar{\kappa}_{gr}$ as a function of $P_g$. This is the *correct* value of $\bar{\kappa}_{gr}$, which was obtained using $T(P_g)$. We follow this procedure for all values of $\bar{\tau}$ and $P_g$. We thus obtain the curve $\bar{\kappa}_{gr}(P_g)$ shown, which gives the correct values of $\bar{\kappa}_{gr}$ in the atmosphere.

Integration yields

$$\frac{1}{n+1} P_g^{n+1} = \frac{g}{\kappa_{gr,0}} \bar{\tau}. \qquad 9.16$$

Here we have used the boundary condition $P_{g0} = 0$.

As long as $H^-$ is responsible for the continuous absorption, $n = 1$ is a good approximation. We then find

$$P_g^2 = \frac{2g}{\kappa_{gr,0}} \bar{\tau} \quad \text{or} \quad P_g = \frac{\sqrt{2}}{\sqrt{\kappa_{gr,0}}} \sqrt{g} \sqrt{\bar{\tau}}. \qquad 9.17$$

Thus, we find that if in the atmosphere $\bar{\kappa}_{gr}$ is approximately proportional to the gas pressure, the gas pressure for a given $\bar{\tau}$ increases proportionally to $\sqrt{g}$. Since for different atmospheres we always see approximately the same $\bar{\tau}$, the pressure in that layer varies approximately as $\sqrt{g}$, and is not proportional to the gravity itself, as one might think at first when looking at the hydrostatic equation. The larger the pressure, the greater is $\bar{\kappa}_{gr}$. Therefore, we see geometrically higher layers in stars with higher gravity (see Fig. 9.4). The net effect is the square root law.

If a larger value of $n$ in equation (9.16) gives a better approximation (for instance, if $n = 2$ is better), then we find an even smaller dependence of the pressure on the gravity, namely $P_g \propto g^{1/3}$.

## 9.4 The electron pressure

Equations (9.10) and (8.30) describe the pressure and temperature stratification of an atmosphere as a function of the mean optical depth $\bar{\tau}$.

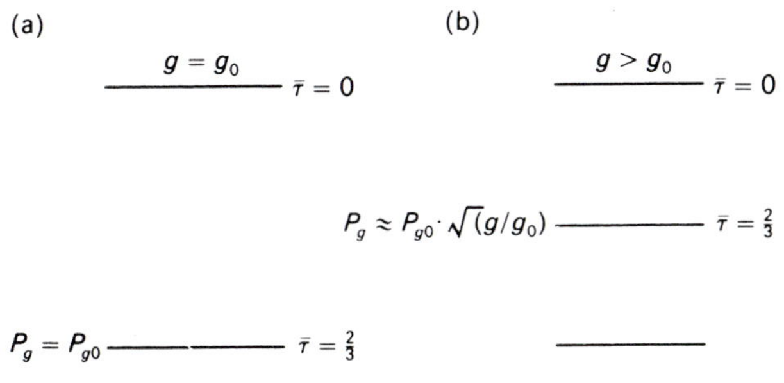

(a)                                          (b)

$g = g_0$                                     $g > g_0$
————————— $\bar{\tau} = 0$                    ————————— $\bar{\tau} = 0$

                         $P_g \approx P_{g0} \cdot \sqrt{(g/g_0)}$ ————— $\bar{\tau} = \frac{2}{3}$

$P_g = P_{g0}$ ——————————— $\bar{\tau} = \frac{2}{3}$            ———————

Fig. 9.4. (a) For an atmosphere with a gravitational acceleration $g = g_0$ we see the layer $\bar{\tau} = \frac{2}{3}$ in which the pressure is equal to $P_{g0}$. (b) In another atmosphere for which the gravitational acceleration is larger than $g_0$, the pressures of all depths are increased. If $\bar{\kappa}$ is proportional to the gas pressure, $\bar{\kappa}$ also increases. We again see the layer where $\bar{\tau} = \frac{2}{3}$, which, however, now corresponds to a smaller geometrical depth, i.e., a higher layer. The pressure at this higher layer is increased only by a factor given by the square root of the gravitational acceleration as compared to the pressure in $\bar{\tau} = \frac{2}{3}$ in the lower gravity atmosphere.

With the gas pressure and the temperature known, we can now obtain the electron pressure which enters into the Saha equation. We can generally say

$$P_g = NkT, \tag{9.18}$$

with $N$ = sum of all particles per $cm^3$.

$$P_e = n_e kT, \tag{9.19}$$

with $n_e$ = number of electrons per $cm^3$. Thus $n_e$ must equal the number of ions + 2 times the number of twice ionized ions + $\cdots$

$$n_e = N^+ + 2N^{2+} + 3N^{3+} + \cdots + iN^{i+}. \tag{9.20}$$

In the simple case of pure hydrogen, we have

$$N = n_e + N(H) + N(H^+) = N(H) + 2n_e \quad \text{since } n_e = N(H^+), \tag{9.21}$$

and for a given $P_g$ we have

$$N = P_g/kT. \tag{9.22}$$

The Saha equation tells us that

$$\frac{N(H^+)n_e}{N(H)} = f(T) \quad \text{or} \quad \frac{N(H^+)^2}{N(H)} = f(T). \tag{9.23}$$

Equations (9.20) and (9.21) yield

$$N(H) + 2N(H^+) = P_g/kT. \tag{9.24}$$

Equations (9.23) and (9.24) are two equations for the two unknowns $N(H^+) = n_e$ and $N(H)$, which can be solved for given $T$ and $P_g$.

If all the hydrogen is ionized, we find, of course, $P_g - P_e = \frac{1}{2}P_g$. If we were to have a pure helium atmosphere for which at very high temperatures all the helium is twice ionized, we would find

$$N = N(He^{2+}) + n_e \quad \text{and} \quad n_e = 2N(He^{2+}), \tag{9.25}$$

which gives

$$\frac{P_g}{P_e} = \frac{3N(He^{2+})}{2N(He^{2+})} = \frac{3}{2}. \tag{9.26}$$

If we have a mixture of several elements, we have to write down the Saha equations for each stage of ionization for each element. We then have to make use of all the equations which describe the abundances of the elements. Equation (9.24) determines the total number of particles and therefore permits the determination of the absolute number of particles per $cm^3$. Equation (9.20) relates the number of electrons to the number of ions. Counting the number of equations and unknowns, we find that we have as many independent equations as we have unknowns. The solution of this system of equations can, however, be quite complicated and is in general done by iteration. Whenever possible, it is easier to give the electron

Table 9.1. *Gas pressures and electron pressures in stars at optical depths* $\bar{\tau} = \frac{2}{3}$ *and* $\bar{\tau} = 0.1$

| $T_{\text{eff}}$ | $\log g$ | $\bar{\tau} = \frac{2}{3}$ | | $\bar{\tau} = 0.1$ | |
|---|---|---|---|---|---|
| | | $\log P_g$ | $\log P_e$ | $\log P_g$ | $\log P_e$ |
| 5500 | 1.0 | 3.08 | −0.04 | 2.75 | −1.07 |
| 5500 | 2.0 | 3.69 | 0.29 | 3.32 | −0.62 |
| 5500 | 3.0 | 4.27 | 0.64 | 3.86 | −0.16 |
| 5500 | 4.0 | 4.83 | 1.01 | 4.42 | 0.32 |
| 5500 | 4.5 | 5.10 | 1.21 | 4.67 | 0.51 |
| 6000 | 2.0 | 3.51 | −0.27 | 3.21 | −0.37 |
| 6000 | 3.0 | 4.15 | 1.04 | 3.83 | 0.02 |
| 6000 | 4.0 | 4.76 | 1.34 | 4.42 | 0.48 |
| 6000 | 4.5 | 5.05 | 1.51 | 4.65 | 0.70 |
| 8000 | 2.0 | 2.37 | 1.58 | 2.15 | 0.68 |
| 8000 | 3.0 | 3.18 | 2.03 | 2.95 | 1.14 |
| 8000 | 4.0 | 3.94 | 2.46 | 3.68 | 1.48 |
| 8000 | 4.5 | 4.31 | 2.65 | 4.04 | 1.64 |
| 10 000 | 2.0 | 1.57 | 1.24 | 1.07 | 0.69 |
| 10 000 | 3.0 | 2.28 | 1.93 | 1.91 | 1.39 |
| 10 000 | 4.0 | 3.03 | 2.59 | 2.73 | 1.93 |
| 10 000 | 4.5 | 3.43 | 2.86 | 3.16 | 2.19 |
| 20 000 | 3.0 | 2.67 | 2.36 | 2.03 | 1.73 |
| 20 000 | 4.0 | 3.40 | 3.09 | 2.75 | 2.45 |
| 20 000 | 4.5 | 3.69 | 3.39 | 3.05 | 2.74 |
| 40 000 | 4.0 | 3.58 | 3.29 | 2.95 | 2.66 |
| 40 000 | 4.5 | 4.04 | 3.75 | 3.45 | 3.16 |

pressure and then calculate the gas pressure, but when we integrate the hydrostatic equilibrium equation we first obtain the gas pressure and then have to calculate the electron pressure.

In Table 9.1 we give gas pressures and electron pressures for stars of different temperatures and different gravitational accelerations for the optical depths $\bar{\tau} = \frac{2}{3}$, which we see in the continuum. We also give the pressures for an optical depth $\bar{\tau} = 0.1$, a typical depth for spectral line formation.

## 9.5 The effects of turbulent pressure

In the gas kinetic derivation of the gas pressure we consider the momentum transfer of a particle to a hypothetical wall in the gas. If a particle has a velocity component $\xi$ in the direction perpendicular to the wall it has a momentum $\vec{M} = m\xi$, where $m$ is its mass, and upon reflection from the wall it has transferred a momentum $\vec{M} = 2m\xi$ to the wall. Here $\xi = v\cos\vartheta$, if $v$ is its velocity and $\vartheta$ the inclination of its velocity with respect to the normal to the surface. The number of particles with the velocity component $\xi$ perpendicular to the wall hitting 1 cm$^2$ each second is given by $N\xi$, where $N$ is the number of particles per cm$^3$. The total momentum transferred per cm$^2$ sec by these particles with the velocity component $\xi$ is then

$$\Delta\vec{M}(\xi) = 2mN\xi^2.\qquad\qquad 9.27$$

The total momentum transfer is obtained by integrating over $\xi$, taking into account the probability for a given $\xi$, which means the Maxwell velocity distribution for one component of the velocity, namely, the one perpendicular to the wall,

$$\varphi(\xi) = \frac{1}{\sqrt{\pi}}\, e^{-(\xi/\xi_{th})^2}\,\frac{d\xi}{\xi_{th}},\qquad\qquad 9.28$$

with $\xi_{th}^2 = 2kT/m$ being the thermal velocity. This leads to

$$P_g = \vec{M} = \frac{2mN}{\sqrt{\pi}}\int_0^\infty \xi^2 e^{-(\xi/\xi_{th})^2}\,\frac{d\xi}{\xi_{th}}.\qquad\qquad 9.29$$

The value of the integral is $\frac{1}{4}\xi_{th}^2\sqrt{\pi}$. With this we obtain

$$P_g = \vec{M} = \frac{2mN\sqrt{\pi}}{\sqrt{\pi}}\frac{2kT}{4}\frac{}{m} = NkT.\qquad\qquad 9.30$$

If, in addition, we have turbulent velocities with an approximately Maxwellian velocity distribution, we find similarly

$$P_t = \vec{M}_t = \frac{2mN}{4}\xi_{turb}^2 = \tfrac{1}{2}\rho\xi_{turb}^2,\qquad\qquad 9.31$$

where $\xi_{turb}$ is now the reference velocity for the turbulent velocity component perpendicular to the hypothetical wall. For a Maxwellian velocity distribution we find

$$\overline{\xi}_{turb}^2 = \tfrac{1}{2}\xi_{turb}^2.\qquad\qquad 9.32$$

We can then also write

$$P_t = \rho\overline{\xi}_{turb}^2.\qquad\qquad 9.33$$

In the presence of turbulent pressure we must now require that the *total* pressure balances the gravitational force. This means

$$\frac{d}{dt}(P_g + P_t) = \frac{dP_g}{dt} + \frac{dP_t}{dt} = g\rho.\qquad\qquad 9.34$$

For the change in gas pressure $P_g$ we thus find

$$\frac{\mathrm{d}P_g}{\mathrm{d}t} = g\rho - \frac{\mathrm{d}P_t}{\mathrm{d}t} = \rho g_{\mathrm{eff}}, \qquad 9.35$$

where

$$g_{\mathrm{eff}} = g - \frac{1}{\rho}\frac{\mathrm{d}P_t}{\mathrm{d}t} = g - \frac{1}{\rho}\frac{\mathrm{d}}{\mathrm{d}t}\,\rho\bar{\xi}^2_{\mathrm{turb}}. \qquad 9.36$$

As long as the turbulent velocity distributions are approximately Maxwellian, we can therefore always take into account a turbulent pressure by replacing $g$ with $g_{\mathrm{eff}}$, as given by equation (9.36).

## 9.6 The effects of radiation pressure

Intense radiation may also have an effect on the pressure stratification. Photons with any energy $h\nu$ have a momentum $\vec{M} = h\nu/c$. When they hit a hypothetical wall they transfer a momentum $(2h\nu\cos\vartheta/c)$ if $\vartheta$ is the angle between the direction of propagation of the photon and the normal to the wall. The number of photons with frequency $\nu$ reaching $1\,\mathrm{cm}^2$ each second is given by

$$\frac{I_\nu}{h\nu}\cos\vartheta\,\mathrm{d}\nu\,\mathrm{d}\omega.$$

The total momentum transfer per $\mathrm{cm}^2$ each second by photons with frequency $\nu$ hitting the wall (or being absorbed) is therefore given by

$$\tfrac{1}{2}\Delta\vec{M} = \int_{\Delta\omega=2\pi}\frac{I_\nu}{c}\cos^2\vartheta\,\mathrm{d}\nu\,\mathrm{d}\omega. \qquad 9.37$$

We now have to add the momentum transfer due to the recoil by the photons emitted or reflected by the wall (or the atoms). The effect of these photons can easily be included by extending the integration in equation (9.37) over the radiation coming back from the wall, which means over the whole solid angle $\mathrm{d}\omega$. This leads to

$$\Delta\vec{M} = \int_0^{2\pi}\int_0^\pi\frac{I_\nu}{c}\cos^2\vartheta\sin\vartheta\,\mathrm{d}\vartheta\,\mathrm{d}\varphi. \qquad 9.38$$

If $I_\nu$ does not depend on $\varphi$, as is to be expected in stars, we find

$$\Delta\vec{M} = \frac{4\pi}{c}K_\nu\,\mathrm{d}\nu \qquad \text{with } K_\nu = -\frac{1}{2}\int_1^{-1}I_\nu\cos^2\vartheta\,\mathrm{d}(\cos\vartheta). \qquad 9.39$$

For the radiation pressure $P_r$ we obtain

$$P_r = \vec{M} = \frac{4\pi}{c}\int_0^\infty K_\nu\,\mathrm{d}\nu = \frac{4\pi}{c}K \approx \frac{4\pi}{3c}J, \qquad 9.40$$

with the Eddington approximation which holds for $\tau \gg 1$, we then find $P_r = \frac{4}{3}\pi J / c$.

For the gradient of the radiation pressure we find

$$\frac{dP_r}{dt} = \frac{4\pi}{c}\int_0^\infty \frac{dK_v}{dt}\,dv = \frac{4\pi}{c}\int_0^\infty \kappa_{v,\text{cm}}\frac{F_v}{4}\,dv, \qquad 9.41$$

according to equation (6.20). Here $\kappa_v$ includes the absorption and scattering coefficients. For the case of a grey atmosphere we can say

$$\frac{dP_r}{dt} = \frac{\bar{\kappa}_{\text{gr}}\rho}{c}\,\pi F_r, \qquad 9.42$$

with $\bar{\kappa} = \bar{\kappa}$ per gram and $F_r =$ radiative flux. Hydrostatic equilibrium then requires

$$\frac{dP_g}{dt} + \frac{dP_t}{dt} + \frac{dP_r}{dt} = g\rho, \qquad 9.43$$

and

$$\frac{dP_g}{dt} = g\rho - \frac{dP_t}{dt} - \frac{dP_r}{dt} = g_{\text{eff}}\,\rho, \qquad 9.44$$

with

$$g_{\text{eff}} = g - \frac{1}{\rho}\frac{dP_t}{dt} - \frac{1}{\rho}\frac{dP_r}{dt} = g - g_t - g_r, \qquad 9.45$$

and

$$g_r = \frac{1}{\rho}\frac{dP_r}{dt} = \frac{\bar{\kappa}_{\text{gr}}}{c}\,\pi F_r. \qquad 9.46$$

If we are not dealing with a grey atmosphere we cannot evaluate the integral in equation (9.41). We are left with the expression

$$g_r = \frac{1}{\rho}\frac{dP_r}{dt} = \frac{1}{c}\int_0^\infty \kappa_{v,\text{gr}}\pi F_v\,dv. \qquad 9.47$$

At large depth, i.e., $\tau_v \gg 1$, we find

$$F_v = \frac{4}{3}\frac{dB_v}{d\tau_v} = \frac{4}{3}\frac{1}{\kappa_{v,\text{cm}}}\frac{dB_v}{dt} \qquad 9.48$$

and we have

$$g_r = \frac{\pi}{c}\frac{1}{\rho}\int_0^\infty \frac{\kappa_{v,\text{cm}}}{\kappa_{v,\text{cm}}}\frac{4}{3}\frac{dB_v}{dt}\,dv = \frac{\pi}{c\rho}\frac{4}{3}\int_0^\infty \frac{dB_v}{dt}\,dv = \frac{\pi}{c\rho}\frac{4}{3}\frac{dB}{dt}. \qquad 9.49$$

with

$$\frac{4\pi}{3}\frac{dB}{dt} = \bar{\kappa}_{\text{cm}}\pi F_r, \qquad 9.50$$

we again obtain $g_r = \bar{\kappa}_{\text{gr}}\pi F_r / v$, even if we are not dealing with a prey atmosphere. For the calculation of the radiative acceleration the non-greyness is only important for $\tau_v < 1$.

Why is only the flux important for the radiative acceleration, which pushes outwards? In an isotropic radiation field there are as many absorption processes from all directions. In each absorption process momentum is transferred to the absorbing atom. The net momentum transfer is therefore zero. Since the emission is isotropic, the net momentum transfer in the emission process is also zero. Only the surplus of absorption processes from one direction is important. The anisotropy of the radiation field is given by $F_v$. Therefore, only $F_v$ counts for the net force outwards.

# 10

---

# Theory of line formation

## 10.1 Formation of optically thin lines

We have seen earlier that we always see a layer with $\tau_\lambda \approx \frac{2}{3}$, i.e., it is always $F_\lambda \approx B_\lambda$ ($\tau_\lambda = \frac{2}{3}$). Spectral lines occur at wavelengths for which $\kappa_\lambda$ is very large. Therefore, we see much higher layers at these wavelengths (Fig. 10.1). These have a lower temperature and so $B_\lambda$ is smaller. This leads to a smaller $F_\lambda$ in the line than $F_c$, the continuum flux in the neighborhood of the line. We define the line depth $R_\lambda$ (see Fig. 10.2) as

$$R_\lambda = \frac{F_c - F_\lambda}{F_c} = 1 - \frac{F_\lambda}{F_c}. \qquad 10.1$$

Clearly, $R_\lambda$ depends on the wavelength within the line. The total area in the line divided by $F_c$ (see the hatched area in Fig. 10.3) is called the *equivalent width* $W_\lambda$. The division by $F_c$ means we are measuring the flux in units of the continuum flux, which means the continuum flux becomes 1. We then have the situation shown in Fig. 10.3: the equivalent width describes the width a rectangular spectral line must have in order to have the same total absorption as the actual line. From this definition we find

$$W_\lambda = \int_{\text{line}} R_\lambda \, d\lambda = \int_{\text{line}} \frac{F_c - F_\lambda}{F_c} \, d\lambda, \qquad 10.2$$

where the integral has to be extended over the whole line. In the following,

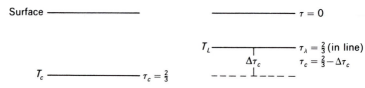

Fig. 10.1. In the wavelength of the line, where $\kappa_\lambda \gg \kappa_c$, we see much higher layers with $T = T_L$ for which $T_L < T_c$ and therefore $F_\lambda = B_\lambda(T_L) < F_c = B_\lambda(T_c)$, where $T_c$ is the temperature of the layer seen in the continuum.

114

$\tau_c$ refers to the optical depth in the continuum, while $\tau_\lambda$ is the optical depth in the line; this means $\tau_\lambda = \tau_{\text{line}} + \tau_c$.

For weak lines, i.e., if $\kappa_{\text{line}} \ll \kappa_{\text{continuum}}$, the layer $\tau_\lambda = \frac{2}{3}$ will be close to the layer with $\tau_c = \frac{2}{3}$; we can make a Taylor expansion of $B_\lambda(T)$ around the layer for which $\tau_c = \frac{2}{3}$ and find

$$B_\lambda(\tau_\lambda = \tfrac{2}{3}) = B_\lambda(\tau_c = \tfrac{2}{3}) - \frac{dB_\lambda}{d\tau_c}\tau_c = \tfrac{2}{3}\Delta\tau_c = B_\tau(\tau_c = \tfrac{2}{3} - \Delta_{\tau_c}). \qquad 10.3$$

It is

$$\frac{\tau_\lambda}{\tau_c} = \frac{\kappa_\lambda}{\kappa_c} \quad \text{or} \quad \tau_c = \tau_\lambda \frac{\kappa_c}{\kappa_c + \kappa_L}, \qquad 10.4$$

where $\kappa_L = \kappa_{\text{line}}$, the line absorption coefficient, and $\kappa_c$ is the continuous absorption coefficient. It is $\kappa_\lambda = \kappa_c + \kappa_L$. For the layer with $\tau_\lambda = \frac{2}{3}$ we find

$$\tau_c = \frac{2}{3}\left(\frac{\kappa_c}{\kappa_c + \kappa_L}\right) = \frac{2}{3} - \Delta\tau_c. \qquad 10.5$$

For $\kappa_L/\kappa_c \ll 1$ the ratio

$$\frac{\kappa_c}{\kappa_c + \kappa_L} \approx 1 - \frac{\kappa_L}{\kappa_c} \quad \text{and} \quad \tau_c \approx \frac{2}{3}\left(1 - \frac{\kappa_L}{\kappa_c}\right) = \frac{2}{3} - \Delta\tau_c$$

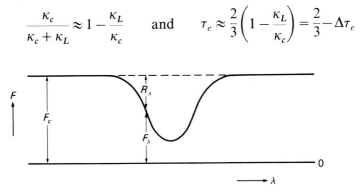

Fig. 10.2. A schematic plot of the energy distribution around a spectral line as a function of wavelength $\lambda$. At the wavelength of the line the flux $F_\lambda$ is smaller than the flux in the continuum $F_c$. $R_\lambda$ is the line depth measured in units of the continuum flux.

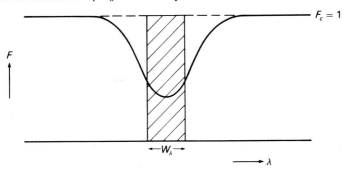

Fig. 10.3. A schematic plot of the energy distribution in the neighborhood of the line. The equivalent width $W_\lambda$ gives the width which a rectangular spectral line must have in order to have the same total absorption as the actual line.

Making use of this, we see that

$$\Delta \tau_c = +\frac{2}{3}\frac{\kappa_L}{\kappa_c} \qquad \text{if } \kappa_L \ll \kappa_c. \qquad\qquad 10.6$$

If $\kappa_L \ll \kappa_c$, the line is called *optically thin*.

With equation (10.6) we find for the line depth

$$R_\lambda = \frac{F_c - F_\lambda}{F_c} = \frac{B_\lambda(\tau_c = \frac{2}{3}) - B_\lambda(\tau_\lambda = \frac{2}{3})}{B_\lambda(\tau_c = \frac{2}{3})}$$

$$= \frac{B_\lambda(\tau_c = \frac{2}{3}) - B_\lambda(\tau_c = \frac{2}{3}) + \frac{dB_\lambda}{d\tau_c}\bigg|_{\tau_c = 2/3}\frac{2}{3}\frac{\kappa_L}{\kappa_c}}{B_\lambda(\tau_c = \frac{2}{3})}, \qquad 10.7$$

or

$$R_\lambda = \frac{2}{3}\frac{\kappa_L}{\kappa_c}\frac{d\ln B_\lambda}{d\tau_c}\bigg|_{\tau_c = 2/3}. \qquad\qquad 10.8$$

For an absorption tube we found (Section 4.4) $I_\lambda = I_{\lambda_0} e^{-\tau_\lambda} \sim I_{\lambda_0}(1 - \tau_\lambda)$, if $\tau_\lambda \ll 1$, and the line depth becomes

$$R_\lambda = \frac{I_{\lambda_0} - I_\lambda}{I_{\lambda_0}} = 1 - (1 - \tau_\lambda) = \tau_\lambda = \kappa_{L,\mathrm{cm}} H_{\mathrm{eff}}, \qquad 10.9$$

where $H_{\mathrm{eff}}$ is the length of the absorption tube.

Comparing equation (10.8) with (10.9), we can write (10.8) in the form

$$R_\lambda = \kappa_{L,\mathrm{cm}} H_{\mathrm{eff}} \qquad \text{with } H_{\mathrm{eff}} = \frac{2}{3}\frac{d\ln B_\lambda}{d\tau_c}\frac{1}{\kappa_{c,\mathrm{cm}}}. \qquad 10.10$$

$H_{\mathrm{eff}}$ can be looked at as the effective height of the atmosphere with respect to line absorption, or in other words, the stellar atmosphere acts as a laboratory absorption tube of length $H_{\mathrm{eff}}$.

Equation (10.8) is a very important relation. It tells us that the line depth is not only determined by the size of the line absorption coefficient, but also by the value for the continuous absorption coefficient and, last but not least, by the gradient of the source function, which can usually be replaced by the gradient of the Planck function. *If there is no temperature gradient with the temperature decreasing outwards, then there are no absorption lines in the spectrum.*

If we want to study the line profile, i.e., the energy distribution within the line, then we have to study the wavelength dependence of $\kappa_L$, as is obvious from equation (10.8).

## 10.2    The line absorption coefficient

The line absorption coefficient $\kappa_L$ per cm$^3$ is given by the absorption coefficient per atom $\kappa_{L,a}$ (dimension cm$^2$) times the number of atoms per cm$^3$ absorbing in the line under consideration, $N_L$. In the case of the Balmer lines, for instance, $N_L$ is given by the number of hydrogen atoms in the second quantum level $N_H$ ($n = 2$). The line absorption coefficient $\kappa_{L,a}$ is not infinitely sharp as one might think, considering that the line is due to a transition of an electron from one bound state to another bound state, for which the frequency of the absorbed or emitted light is given by

$$h\nu = \chi \text{ (upper state)} - \chi \text{ (lower state)},$$

where $\chi$ is the excitation energy. $\kappa_{L,a}$ extends over a certain wavelength band. The reason for this natural width of the line is the natural width of the energy levels. In the classical approach of the oscillating electron, the emission in the line is the emission of a bound oscillating electron with eigenfrequency $\nu$. Due to this radiation, the oscillation is damped and, as we saw before, the emitted energy has a frequency dependence given by the emissivity $\varepsilon_{L,a}$ (since $\varepsilon_{L,a} = \kappa_{L,a}S_\nu = \kappa_{L,a}B_\nu$ in LTE, the profile of $\kappa_{L,a}$ resembles that of $\varepsilon_{L,a}$). With $\omega_0$ the frequency in the center of the line, one finds

$$\varepsilon_{L,a} \propto \frac{\gamma}{(\omega^2 - \omega_0^2)^2 + \gamma^2\omega^2} \sim \frac{\gamma}{(2\omega)^2(\omega - \omega_0)^2 + \gamma^2\omega^2} = \frac{\gamma}{(\Delta\omega^2 + \gamma^2/4)} \frac{1}{2^2\omega^2}.$$

10.11

Here we have used

$$\omega^2 - \omega_0^2 = (\omega - \omega_0)(\omega + \omega_0) \sim 2\omega(\omega - \omega_0) \qquad \text{for} \quad \frac{\omega - \omega_0}{\omega} \ll 1.$$

The frequency $\omega = 2\pi\nu$ can be considered to be constant over the line. The important quantity for the line profile is the denominator. We find

$$\kappa_{L,a} \propto \frac{\gamma}{\Delta\omega^2 + (\gamma/2)^2}.$$

10.12

This so-called *damping profile*, or *Lorentzian profile*, has a total half width $2\Delta\omega(\tfrac{1}{2}) = \gamma$ (see Fig. 10.4). The larger the amount of radiation, the larger and the broader the line.

From the classical damping theory of the oscillating and radiating electron one finds

$$\gamma = \frac{8\pi^2 e^2}{3mc\lambda_0^2},$$

10.13

where $\lambda_0$ is the center wavelength of the line.

The natural total half width of the line for a classical electron comes out to be $\lambda_N = 1.2 \times 10^{-4}$ Å, independently of the wavelength $\lambda$.

For the absorption coefficient $\kappa_{L,a}$ we find

$$\kappa_{L,a} = \frac{\pi e^2}{mc} \frac{\gamma}{\Delta\omega^2 + (\gamma/2)^2} f. \qquad 10.14$$

Here we have introduced the factor $f$, which is called the *oscillator strength* because this factor indicates how many classical oscillators the line under consideration corresponds to.

In the quantum-mechanical approach, the natural width of the line is determined by the finite width of the two energy levels. According to the Heisenberg uncertainty principle, the uncertainty of an energy measurement $\Delta E$ is given by the time $\Delta t$ which is available for the measurement, namely

$$\Delta E \, \Delta t = \hbar = h/2\pi, \qquad h = \text{Planck's constant}. \qquad 10.15$$

The maximum time available for the measurement of the energy $E$ is given by the time which the electron stays in that energy level. The larger the transition probability, the shorter the lifetime, the stronger the radiation and the larger $\Delta E$, which means the larger the natural line width. It turns out then that $\gamma$ is proportional to the transition probability. This means that the natural line width is of the order

$$\Delta\omega/\omega = \Delta\lambda/\lambda \propto \text{transition probability}. \qquad 10.16$$

Actually, it is the sum over all the transition probabilities for the level under consideration. When calculating the line width we obviously have to take into account the widths of both levels which are involved in the line transition.

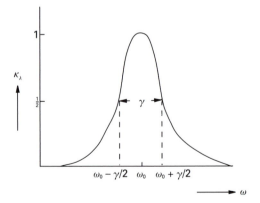

*Fig. 10.4.* A schematic plot of the classical damping profile for an oscillating electron with eigenfrequency $\omega_0$ and a damping constant $\gamma$.

In most astrophysical cases the line width is, however, determined not by the radiation damping, but by the thermal motion of the emitting and absorbing atoms.

## 10.3   The Doppler profile

The light emitting atoms in the stellar atmosphere are not at rest, but have a Maxwell velocity distribution which for one velocity component $\xi = v_x$ is given by

$$\frac{dN}{N} = \frac{1}{\sqrt{\pi}} e^{-(\xi/\xi_0)^2} \frac{d\xi}{\xi_0} \qquad \text{where } \zeta_0^2 = \frac{2R_g T}{\mu}. \qquad 10.17$$

A particle moving with $\xi$ away from the observer (see Fig. 10.5) emits light at a wavelength $\omega_0 + \Delta\omega$ with

$$\frac{\Delta\omega}{\omega_0} = -\frac{\xi}{c} \qquad \text{or} \qquad \frac{\Delta\omega}{\omega_0} = \frac{\xi}{c} \qquad 10.18$$

for motion away from the observer      for motion towards the observer

Only the component of the motion in the direction of the line of sight has to be considered for the Doppler effect. Therefore, we have to consider the Maxwell distribution for one velocity component only.

The number of emission or absorption processes at $\omega + \xi\omega/c$ is then given by the number of atoms moving with a velocity component $\xi$ towards us. This means that the emission or absorption profile must be given by

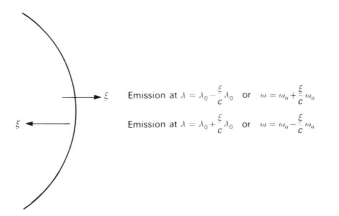

*Fig. 10.5.* Thermal motions of amount $\xi$ of the light emitting particles leads to emission at frequencies $\omega = \omega_0 \pm \omega_0\xi/c$ seen by us, the observers. $\omega = \omega_0$ in the rest frame of the particle.

the Maxwell distribution (10.17), or

$$\kappa_{L,a} \propto N(\zeta) \propto e^{-(\xi/\xi_0)^2}\frac{1}{\xi_0} \propto e^{-(\Delta\omega/\Delta\omega_D)^2}\frac{1}{\Delta\omega_D}, \qquad 10.19$$

with

$$\frac{\xi}{c} = \frac{\Delta\lambda}{\lambda} = -\frac{\Delta\omega}{\omega} \quad \text{and} \quad \frac{\xi_0}{c} = \frac{\Delta\lambda_D}{\lambda} = -\frac{\Delta\omega_D}{\omega} \quad \text{and} \quad \xi_0 = \sqrt{\left(\frac{2R_gT}{\mu}\right)},$$

or

$$\kappa_{L,a} \propto e^{-(\Delta\lambda/\Delta\lambda_D)^2}\frac{1}{\Delta\lambda_D} \quad \text{with } \Delta\lambda_D = \lambda\frac{\xi_0}{c}. \qquad 10.20$$

For an electron which would actually radiate like a classical oscillator, one obtains

$$\int_{-\infty}^{\infty}\kappa_\nu \, d\Delta\nu = \frac{\pi e^2}{mc} \quad \text{or} \quad \int_{-\infty}^{\infty}\kappa_\lambda \, d\Delta\lambda = \frac{\lambda^2}{c}\frac{\pi e^2}{mc}, \quad \text{since } d\lambda = -\frac{c}{\nu^2}\,d\nu.$$

$$10.21$$

With the classical radius of the electron being given by

$$r_{\text{el}} = \frac{e^2}{mc^2},$$

we can also write

$$\int_{-\infty}^{\infty}\kappa_\lambda \, d\Delta\lambda = \lambda^2\pi r_{\text{el}}.$$

The integral over $\kappa_\lambda$ over the line must, of course, remain the same, independently of the Doppler broadening. In a frequency independent radiation field the amount of energy absorbed by the atoms does not change when the atoms are in motion. The condition (10.21) then leads to the value for $\kappa_{L,a}$:

$$\kappa_{L,a} = \frac{\sqrt{\pi}e^2}{mc^2}\frac{\lambda^2}{\Delta\lambda_D}e^{-(\Delta\lambda/\Delta\lambda_D)^2}. \qquad 10.22$$

Actually, the radiation for electrons in different energy states and for different transitions is different in intensity, which is not obvious from equation (10.21). As mentioned earlier, one therefore introduces a factor $f$ describing the actual strength of the lines. This factor tells us to how many classical oscillators the emission corresponds. Therefore, we call this factor the oscillator strength of the line, or the transition.

For broadening by the Doppler effect only, we find a half width

$$\Delta\lambda_{1/2} = \Delta\lambda_D\sqrt{(\ln 2)}. \qquad 10.23$$

For the sun with $T \sim 6000$ K we find for an iron atom with atomic weight $\mu = 56$ that

$$\xi_0 = \sqrt{\left(\frac{2RT}{\mu}\right)} = \sqrt{\left(\frac{2 \times 8.3 \times 10^7 \times 6 \times 10^3}{56}\right)} \approx 1.4 \times 10^5 \text{ cm s}^{-1},$$

or

$$\xi_0 = 1.4 \text{ km s}^{-1} \quad \text{and} \quad \frac{\Delta\lambda_D}{\lambda} = \frac{1.4 \times 10^5}{3 \times 10^{10}} \sim 4 \times 10^{-6}.$$

For $\lambda \sim 4000$ Å the $\Delta\lambda_D \sim 1.6 \times 10^{-10} \sim 0.016$ Å, i.e., the Doppler width is much larger than the damping width of the line, which is about $1.1 \times 10^{-4}$ Å, as we saw above.

For the hydrogen atom at $T = 6000$ K, we find

$$\xi_0 = \sqrt{\left(\frac{2 \times 8.3 \times 10^7 \times 6 \times 10^3}{1}\right)} \approx 10 \text{ km s}^{-1}.$$

This is always a good reference point to estimate average thermal velocities. The Doppler width of the Balmer lines is about 0.1 Å. However, in most of the stars the hydrogen lines are much wider because of Stark effect broadening, which we shall discuss in Chapter 11.

## 10.4   The Voigt profile

Generally, we have to consider both effects: the Doppler profile and the damping profile. The damping profile is negligible in the line center because the Doppler width is so much larger, but in the wings the Doppler profile decreases very steeply, while the damping profile decreases only as $1/\Delta\lambda^2$. The general profile, also called the '*Voigt*' profile, consists therefore of a Doppler core and damping wings (see Fig. 10.6). The final form of the profile depends on the ratio of the damping widths $\gamma/2$ to the Doppler width $\Delta\omega_D$. This ratio is usually denoted by $\alpha$, i.e., $\alpha = \gamma/2\Delta\omega_D$. The line absorption coefficient $\kappa_L$ is then given by

$$\kappa_{L,a} = \kappa_0 H(\alpha, v) \quad \text{with} \quad \kappa_0 = \frac{\sqrt{\pi} \cdot e^2}{mc^2} \frac{\lambda^2}{\Delta\lambda_D} \cdot f. \qquad 10.24$$

In Table 10.1 we tabulate the function from which the Voigt profiles can be calculated.

In Fig. 10.7 we show a logarithmic plot of line profiles for different values of $\alpha$. As a rule of thumb we can say that we start to see the damping wings at a distance $\Delta\lambda$ from the line center for which

$$\Delta\lambda = -(\log \alpha)\Delta\lambda_D. \qquad 10.25$$

This rule is often useful for a first guess of the Doppler widths $\Delta\lambda_D$.

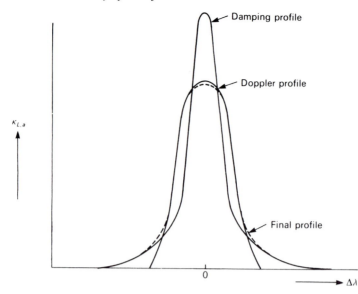

Fig. 10.6. A schematic plot of the different line profiles contributing to the Voigt profile. Generally, the line profile consists of the Doppler core and the damping wings. It is called a Voigt profile.

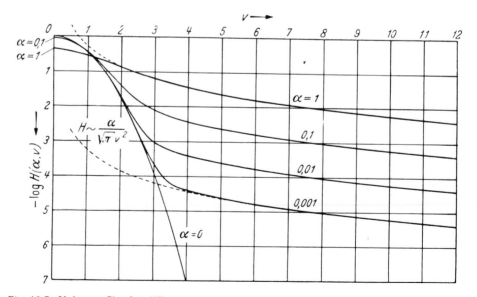

Fig. 10.7. Voigt profiles for different values of $\alpha$ are shown in a logarithmic plot. (From Unsöld, 1955, p. 265.)

Table 10.1. *Functions needed to calculate the Voigt function* $H(\alpha, v)$ *for* $\alpha \ll 1$ *according to D. L. Harris (1948)*

$$v = \frac{\Delta\lambda}{\Delta\lambda_D} = \frac{\Delta\omega}{\Delta\omega_D} \qquad \alpha = \frac{\gamma}{2\Delta\omega_D}$$

$$H(\alpha, v) = e^{-v^2} - \frac{2\alpha}{\sqrt{\pi}}[1 - 2vF(v)] \quad \text{with} \quad F(v) = e^{-v^2}\int_0^v e^{u^2}\,du$$

| $v$ | $e^{-v^2}$ | $-\dfrac{2}{\sqrt{\pi}}[1 - 2vF(v)]$ |
|---|---|---|
| 0.0 | +1.000 000 | −1.128 38 |
| 0.1 | +0.990 050 | −1.105 96 |
| 0.2 | +0.960 789 | −1.040 48 |
| 0.3 | +0.913 931 | −0.937 03 |
| 0.4 | +0.852 144 | −0.803 46 |
| 0.5 | +0.778 801 | −0.649 45 |
| 0.6 | +0.697 676 | −0.485 52 |
| 0.7 | +0.612 626 | −0.321 92 |
| 0.8 | +0.527 292 | −0.167 72 |
| 0.9 | +0.444 858 | −0.030 12 |
| 1.0 | +0.367 879 | +0.085 94 |
| 1.1 | +0.298 197 | +0.177 89 |
| 1.2 | +0.236 928 | +0.245 37 |
| 1.3 | +0.184 520 | +0.289 81 |
| 1.4 | +0.140 858 | +0.313 94 |
| 1.5 | +0.105 399 | +0.321 30 |
| 1.6 | +0.077 305 | +0.315 73 |
| 1.7 | +0.055 576 | +0.300 94 |
| 1.8 | +0.039 164 | +0.280 27 |
| 1.9 | +0.027 052 | +0.256 48 |
| 2.0 | +0.018 3156 | +0.231 726 |
| 2.1 | +0.012 1552 | +0.207 528 |
| 2.2 | +0.007 9071 | +0.184 882 |
| 2.3 | +0.005 0418 | +0.164 341 |
| 2.4 | +0.003 1511 | +0.146 128 |
| 2.5 | +0.001 9305 | +0.130 236 |
| 2.6 | +0.001 1592 | +0.116 515 |
| 2.7 | +0.000 6823 | +0.104 739 |
| 2.8 | +0.000 3937 | +0.094 653 |
| 2.9 | +0.000 2226 | +0.086 005 |
| 3.0 | +0.000 1234 | +0.078 565 |
| 3.1 | +0.000 0671 | +0.072 129 |
| 3.2 | +0.000 0357 | +0.066 526 |
| 3.3 | +0.000 0186 | +0.061 615 |
| 3.4 | +0.000 0095 | +0.057 281 |

Table 10.1 *continued*

| $v$ | $e^{-v^2}$ | $-\dfrac{2}{\sqrt{\pi}}[1 - 2vF(v)]$ |
|---|---|---|
| 3.5 | +0.000 0048 | +0.053 430 |
| 3.6 | +0.000 0024 | +0.049 988 |
| 3.7 | +0.000 0011 | +0.046 894 |
| 3.8 | +0.000 0005 | +0.044 098 |
| 3.9 | +0.000 0002 | +0.041 561 |
| 4.0 | +0.000 0001 | +0.039 250 |
| 4.2 | — | +0.035 195 |
| 4.4 | — | +0.031 762 |
| 4.6 | — | +0.028 824 |
| 4.8 | — | +0.026 288 |
| 5.0 | — | +0.024 081 |

## 10.5    Line broadening due to turbulent motions

The absorption or emission of radiation by the atoms is, of course, also Doppler shifted if other motions are present in addition to the thermal motions in the atmosphere. We saw, for instance, the Doppler shifts due to the solar granulation motion, as seen in the wiggly lines of the solar spectrum (see Volume 1, Chapter 18). If the sun were very far away and we could only take a spectrum of the light integrated over the whole sun the lines would no longer be wiggly, but would be correspondingly broadened. We would call this *macroturbulent broadening*, meaning that *different parts of the stellar disc have different velocities*. If we assume that these macroturbulent velocities have a velocity distribution similar to the distribution of the thermal velocities, i.e., Gaussian, then the resulting velocity distribution obtained after folding the macroturbulent velocity distribution with the thermal velocity distribution is again a Gaussian distribution with a reference velocity $\xi_0$, which is given by

$$\xi_0^2 = \xi_{\mathrm{th}}^2 + \xi_{\mathrm{turb}}^2, \qquad\qquad 10.26$$

and the Doppler width of the line then corresponds to $\Delta\lambda_D$, which is now determined by the $\xi_0$ given in equation (10.26). $\xi_{\mathrm{th}}$ stands for the reference velocity of thermal motions and $\xi_{\mathrm{turb}}$ is the reference velocity of the turbulent motions.

In addition to the macroturbulent motions, we may also have so-called *microturbulence*. If along the line of sight down to $\tau_\lambda = \frac{2}{3}$ we encounter different velocities, then the photons coming out of the atmosphere encounter atoms with different velocities, just as in the case of the thermal motions. In this case it appears to the photons as if the absorption coefficient $\kappa_{L,a}$ is broadened. It appears as if the thermal motions are enhanced. The lines are broadened just as in the case of macroturbulence, but the effect on the total line absorption in a given line is different than in the case of macroturbulence. In the case of macroturbulence, think of the solar granulation, each part of the stellar disc is independent of the others. For instance, a line in a granulum becomes optically thick for a given number of absorbing atoms, no matter whether the matter in the granulum is moving up or down or whether it is at rest. Within each granulum the line profile and the total absorption does not change. It can easily be seen that when you add up all the spectra for the many granules the total absorption does not change either. Only the apparent line profile becomes wider because the lines from the different granules appear at different wavelengths. This is different for microturbulence. Since along the line of sight the atoms now have different velocities, a photon with a small wavelength shift with respect to the center of the line which would have been reabsorbed on the way out will now possibly not be reabsorbed again because the material on top may be moving out very fast, and the absorption coefficient at the wavelength of the photon is now very small. At any given wavelength within the line the absorption coefficient appears to be altered. The $\kappa_L/\kappa_c$ at each wavelength within the line is changed. For an increasing number of absorbing atoms the line remains optically thin much longer than if no microturbulence were present. This has important consequences when we look at the total absorption in such lines.

## 10.6 Other distortions of the line profiles

Another intrinsic distortion of the line profiles in stars is due to stellar rotation, as we discussed in Volume 1, Chapter 13.

As demonstrated in Fig. 10.8, one half of the star moves away from us, while the other half moves towards us. The component of motion in the direction of the line of sight is $v_y = v_r(\vartheta) \sin \phi$, if $v_r(\vartheta)$ is the rotational velocity for the colatitude $\vartheta$. For rigid body rotation $v_r(\vartheta) = v_e \sin \vartheta$, and the velocity component $v_y$ along the line of sight is $v_y = v_e \sin \phi \sin \vartheta$, where $v_e$ is the equatorial rotational velocity. $v_y$ is constant for all points on the sphere for which $\sin \phi \sin \vartheta = \text{const}$. Where these points are on the sphere can best be seen if we describe the sphere of the star in spherical polar coordinates. It is (see Fig. 10.9):

$$x = \rho \sin \phi = r \sin \vartheta \sin \phi$$
$$y = \rho \cos \phi = r \sin \vartheta \cos \phi$$
$$z = r \cos \vartheta.$$

This shows that $\sin \vartheta \sin \phi$ is constant for $x = $ const., i.e., for stripes across the star (see Fig. 10.10). A spectral line broadened by rotation only has an elliptical line profile (see Fig. 10.11) if all areas on the stellar surface are equally bright.

The spectral line profile is generally also distorted by the finite resolution of the spectrograph, which gives a line profile a certain shape, even if the spectral line were intrinsically infinitely sharp. This line profile is called the

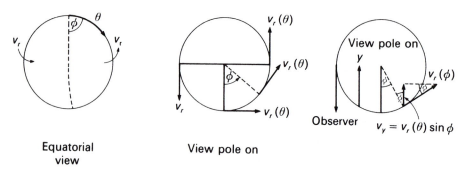

Equatorial view          View pole on

Fig. 10.8. One half of a rotating star moves towards us. The other half moves away from us.

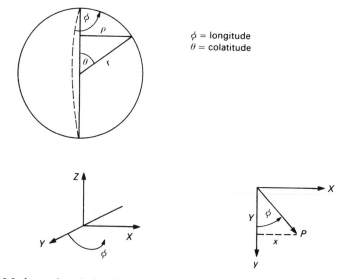

$\phi = $ longitude
$\theta = $ colatitude

Fig. 10.9 shows the relation between the cartesian coordinates $X, Y, Z$ and the spherical polar coordinates, $r, \vartheta, \phi$.

*instrumental profile*. It gives an additional line broadening, especially for low dispersion spectra. For such low dispersion spectra the instrumental broadening is frequently more important than any broadenings intrinsic to the star. On such low dispersion spectra we cannot therefore measure intrinsic line profiles. For high dispersion spectra we have to correct the measured line profile for the instrumental broadening, which has to be measured in the laboratory. Quite often we are unable to extract the true line profiles, especially for spectra of faint stars, for which we can only get low resolution spectra. We can then only work with the total absorption in the lines. The total absorption in the lines remains unchanged due to all these macroscopic broadening mechanisms. (This does not prevent us

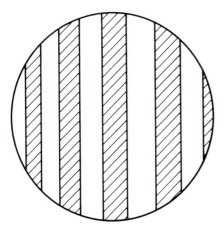

Fig. 10.10. $v_y$ = const. for stripes on the star's disc parallel to the projected axis of rotation.

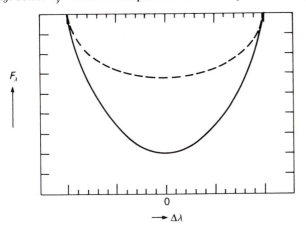

Fig. 10.11. A spectral line broadened by rotation only has an elliptical line profile for a star with no limb darkening (---). (——) profile with strong limb darkening. (From Gray, 1976, p. 398.)

from measuring slightly wrong values for lower dispersion spectra because we tend to make wrong judgements for the position of the continuum on such low dispersion spectra.)

## 10.7    Equivalent widths for optically thin lines

The equivalent width for an optically thin line can now easily be calculated, making use of equations (10.2) and (10.8). We find

$$
\begin{aligned}
W_\lambda = \int_{\text{line}} R_\lambda \, d\lambda &= \frac{2}{3} \frac{1}{\kappa_c} \frac{d \ln B_\lambda}{d\tau_c} \int_{\text{line}} \kappa_{\text{line}} \, d\lambda \\
&= \frac{2}{3} \frac{1}{\kappa_c} \frac{d \ln B_\lambda}{d\tau_c} N_{rs} \int_{\text{line}} \kappa_{\text{line}} \text{ (per atom) } d\lambda \\
&= C N_{rs} \lambda^2 \frac{\pi e^2}{mc^2} f_{rs} = C' \lambda^2 N_{rs} f_{rs}.
\end{aligned}
\tag{10.27}
$$

Here $r$ is the index indicating the degree of ionization of the absorbing ion or atom and the second index $s$ describes the quantum numbers of the two levels involved in the transition. Equation (10.27) tells us that for optically thin lines the equivalent width is proportional to the number of absorbing atoms times the oscillator strength. With

$$
N_{rs} = N_{ro} \frac{g_{rs}}{g_{ro}} e^{-\chi_s / kT},
$$

we find that

$$
W_\lambda \propto \frac{N_{ro}}{g_{ro}} g_{rs} f_{rs} \lambda^2 e^{-\chi_s / kT}.
\tag{10.28}
$$

If we measure $W_\lambda$ and $g_{rs} f_{rs}$ in the laboratory as well as $\chi_s$, we can determine $N_{ro}/g_{ro}$ and thereby the number of atoms or ions of the particular element for which the line was observed. However, in order to make use of the Boltzmann formula we must know the temperature. We have already discussed several methods for the determination of atmospheric temperatures. In this context it seems to be best to determine the temperature directly from the study of lines which originate from different levels with different excitation energies. If we measure the $W_\lambda$ for two such lines, one with index 1 and another with index 2, we find for the equivalent widths

$$
W_{\lambda_1} \propto \frac{N_{ro}}{g_{ro}} g_{r_1} f_1 \lambda_1^2 e^{-\chi_1 / kT}
$$

$$
W_{\lambda_2} \propto \frac{N_{ro}}{g_{ro}} g_{r_2} f_2 \lambda_2^2 e^{-\chi_2 / kT}.
\tag{10.29}
$$

The ratio of the equivalent widths gives

$$\frac{W_{\lambda_1}}{W_{\lambda_2}} = \frac{g_{r_1} f_1}{g_{r_2} f_2} \frac{\lambda_1^2}{\lambda_2^2} e^{-(\chi_1 - \chi_2)/kT}. \tag{10.30}$$

If the excitation energies are known, the temperature can be determined. This so determined temperature is called the *excitation temperature* because it is determined from the excitation difference of the two levels. We can also say that we determine $T$ in such a way that both lines give the same $N_{ro}/g_{ro}$, as must be the case.

All these derivations only hold as long as $\kappa_{\text{line}}/\kappa_c \ll 1$. For most spectral lines observed this is not true. In fact, all lines with $W_\lambda \geqslant 15$ mÅ are not strictly optically thin lines.

## 10.8 Optically thick lines

Equation (10.8) tells us the shape of the line profile in the optically thin case. How does the line look when the line becomes optically thick? Fig. 10.12 demonstrates schematically the line profiles for increasing values of $\kappa_L/\kappa_c$ or with increasing number of absorbing atoms. As long as $\kappa_L/\kappa_c \ll 1$, i.e., for optically thin lines, the line depth at every point within the lines increases proportional to $\kappa_L/\kappa_c$, or proportional to the number of atoms $N_s$ absorbing in the line. This is no longer the case when $\kappa_L/\kappa_c \approx 1$ or greater than 1. In this case, the lines are called optically thick. When lines become optically thick they necessarily reach a maximum depth $R_\lambda$, because $R_\lambda$ cannot become larger than 1 (we cannot have less than zero intensity). Actually, the lowest intensity in the line center is given by $S_\lambda(\tau_\lambda = \frac{2}{3})$ in the line center. For optically very thick lines for which $\kappa_L/\kappa_c$ in the line center goes to infinity, the intensity in the line center is given by the source function $S_\lambda(\tau_c = 0)$. The lowest central intensity in the line is therefore given by $S_\lambda(0)$ or in LTE by $B_\lambda(0)$ which is not zero, since $T(0)$ is not zero (see equation

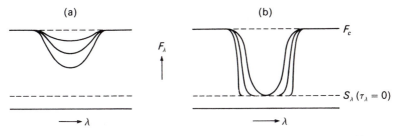

*Fig. 10.12.* Changes of the line profile with increasing $\kappa_L/\kappa_c$ for (a) optically thin and (b) optically thick lines.

8.30). (In cases of NLTE when $S_\lambda \neq B_\lambda$ the $S_\lambda(0)$ may tend towards zero, for instance, in resonance lines, but these are exceptional cases, although usually the strongest lines.)

In the wings the line is still optically thin. There the line depth still grows proportionally to $\kappa_L/\kappa_c$, so the line becomes wider, and the profile becomes more rectangular. Since the line flanks are very steep, the increase in total absorption is very small with increasing values of $\kappa_L/\kappa_c$. The increase in optical thickness thus gives an additional line broadening which might be mistaken for an additional broadening due to turbulence, especially if the line is broadened by rotation or by the instrumental profile such that the line depth may not appear to be very deep (see, for instance, Volume 1, Fig. 13.3, where we show an optically very thick line in a non-rotating star and in a star with high rotational velocity). It appears similar to an optically thin line with very high turbulent velocities. It is not always easy to distinguish between these possibilities. The best way seems to be by Fourier analysis, although extremely accurate measurements are necessary.

## 10.9    The curve of growth

### 10.9.1    *Interpolation of the line profile*

In this section we will discuss how the total absorption within the spectral line, measured by the equivalent width $W_\lambda$ (see Section 10.1), is related to the line absorption coefficient $\kappa_L$ and thereby to the number of atoms absorbing in a given line. Once we understand this relation we will be able to determine the number of absorbing atoms, and from that determine the chemical composition of the stars.

The total absorption in the line is given by

$$W_\lambda = \int_{\text{line}} \frac{F_c - F_\lambda}{F_c} \, d\lambda = \int_{\text{line}} R_\lambda \, d\lambda, \qquad 10.31$$

where the integral has to be extended over the whole line. In the previous sections we have discussed the wavelength dependence of $R_\lambda$ for optically thin lines. We have also seen that for optically thick lines the limiting central line depth is given by $R_{\lambda_0}$ with

$$R_{\lambda_0} = \frac{F_c - S_\lambda(0)}{F_c}. \qquad 10.32$$

In order to describe the whole line profile, we need an interpolation for $R_\lambda$ which is equal to equation (10.8) in the part of the line which is optically thin, i.e., in the wings of the line, and which equals equation (10.32) in the optically thick parts of the line. Such an interpolation was given by Minnaert. We introduce a new quantity

$$X(\lambda) = \frac{2}{3} \frac{\kappa_L(\lambda)}{\kappa_c} \frac{d \ln B_\lambda}{d\tau_c}, \qquad 10.33$$

with the derivative taken at $\tau_c = \frac{2}{3}$, and with $\kappa_L(\lambda_0) = \dfrac{\sqrt{\pi}e^2}{mc^2} \dfrac{\lambda_0^2}{\Delta\lambda_D} N_s f$, and

$N_s$ = number of absorbing atoms per cm³.

According to equation (10.8) we find for optically thin lines that $R_\lambda = X(\lambda)$. We also see that in the line center of optically thick lines where $R_\lambda$ is given by equation (10.32), the $R_{\lambda_0}$ is much smaller than $X(\lambda_0)$. The quantity $X(\lambda)$ can obviously become much larger than 1, while $R_\lambda$ cannot. So for

$$X(\lambda) \gg 1 \qquad \text{we find} \qquad R_\lambda(\lambda) \ll X(\lambda).$$

With this in mind, we can now obtain a very good approximation for the line profile by the interpolation formula

$$R_\lambda = \frac{1}{\dfrac{1}{X(\lambda)} + \dfrac{1}{R_{\lambda_0}}}. \qquad 10.34$$

For very large values of $\kappa_L/\kappa_c$ the $X(\lambda)$ is much larger than 1. It is then also much larger than $R_{\lambda_0}$ such that $1/X(\lambda)$ is much smaller than $1/R_{\lambda_0}$ and we have $R_\lambda = R_{\lambda_0}$.

For small $X(\lambda)$, i.e., optically thin lines, $1/X(\lambda) \gg 1/R_{\lambda_0}$ and $R_\lambda \sim X(\lambda)$. As we saw before for a Doppler profile

$$X(\lambda) = X(\lambda_0 + \Delta\lambda) = X(\lambda_0)e^{-(\Delta\lambda/\Delta\lambda_D)^2}, \qquad 10.35$$

with

$$X(\lambda_0) = \frac{\sqrt{\pi}e^2}{mc^2} \frac{\lambda_0^2}{\Delta\lambda_D} N_s f H_{\text{eff}}. \qquad 10.36$$

The integral

$$W_\lambda = \int_{\text{line}} R_\lambda \, d\lambda = \int_{\text{line}} \frac{1}{\dfrac{1}{X(\lambda)} + \dfrac{1}{R_{\lambda_0}}} \, d\lambda$$

can be computed. What we find is a relation between $W_\lambda$ and $X(\lambda_0)$, $\Delta\lambda_D$, and $R_{\lambda_0}$.

The relation between $W_\lambda$ and $X(\lambda_0)$ is called the *curve of growth*. For a Doppler profile it depends on the two parameters $R_{\lambda_0}$ and $\Delta\lambda_D$.

### 10.9.2    *The curve of growth*

The curve of growth describes how the equivalent width $W_\lambda$ depends on $X(\lambda_0)$. Since $X(\lambda_0) \propto N_s$, the number of absorbing atoms, this is equivalent to saying how $W_\lambda$ depends on the number of atoms absorbing in the lines for a given $\kappa_c$ and a given $\mathrm{d}\ln B_\lambda/\mathrm{d}\tau_c$.

For all optically thin lines, we saw above that the line depth at any point within the line grows proportionally to $\kappa_L/\kappa_c$, or proportionally to $N_s$. The integral over the line depth, which is equal to $W_\lambda$, therefore also grows proportionally to $N_s$. For large $X(\lambda_0)$ the growth of $W_\lambda$ becomes much smaller, since no absorption is added any more in the line center. The line only increases very slowly in width. The equivalent width therefore increases only very slowly. We find

$$W_\lambda \propto \sqrt{(\ln X(\lambda_0))} \propto \sqrt{(\ln N_s f)}.  \qquad 10.37$$

This part of the curve is called the *flat part* of the curve of growth (see Fig. 10.13). If $X(\lambda_0)$ becomes very large, the line becomes so strong that even the weak, far out wings due to the damping become visible. Then the equivalent width increases again and one finds

$$W_\lambda \propto \sqrt{(N_s \gamma f)},  \qquad 10.38$$

where $\gamma$ is the damping constant (see Section 10.2).

This part is called the *damping part*, or the *square root part*, of the curve of growth.

The extent of the flat part of the curve of growth depends very much on the ratio $\gamma/(2\Delta\lambda_D) = \alpha$, as seen in Fig. 10.13, which shows schematic curves of growth.

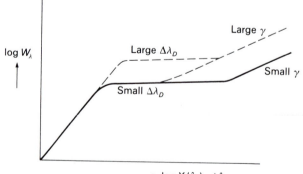

Fig. 10.13. Schematic curve of growth for different values of the Doppler width $\Delta\lambda_D$ and damping constants $\gamma$.

### 10.9.3 *The influence of the Doppler width on the curve of growth*

We still have to discuss how the Doppler width enters into the curve of growth.

Fig. 10.14 shows that for increasing line width saturation occurs for larger $W_\lambda$; in fact, for the same degree of saturation or the same depth in the line center the equivalent width increases proportionally to the line width, i.e., proportionally to $\Delta\lambda_D$: so the flat part of the curve of growth is higher by a factor proportional to $\Delta\lambda_D$. This is demonstrated in Figs. 10.13 and 10.14. However, if in the curve of growth we plot $\log(W_\lambda/\Delta\lambda_D)$, then the level of the flat part of the curve of growth remains independent of $\Delta\lambda_D$. In order to keep the linear part of the curve of growth in place, we have to divide the abscissa by $\Delta\lambda_D$ as well and plot $\log(W_\lambda/\Delta\lambda_D)$ as a function of $\log(X(\lambda_0)$. We then obtain one curve of growth for all values of $\Delta\lambda_D$ (except for the damping part), which is quite useful if we want to do spectrum analysis by means of the curve of growth.

### 10.9.4 *The influence of $S_\lambda(0)$ on the curve of growth*

From Fig. 10.15 it is obvious that for very strong lines the flat part of the curve of growth will increase proportionally to the central line depth $R_{\lambda_0}$, since the line profile is almost rectangular. Plotting $\log(W_\lambda/R_{\lambda_0}\Delta\lambda_D)$ as a function of $\log(X(\lambda_0)/R_{\lambda_0})$ will give us a curve of growth which is independent of $R_{\lambda_0}$ and $\Delta\lambda_D$.

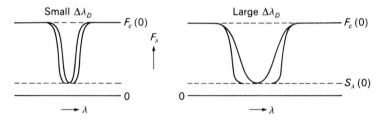

Fig. 10.14. For the same central line depth, the equivalent widths of optically thick lines increase proportionately to the line widths.

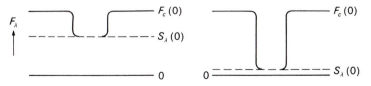

Fig. 10.15. For saturated lines the equivalent width becomes proportional to the central line depth.

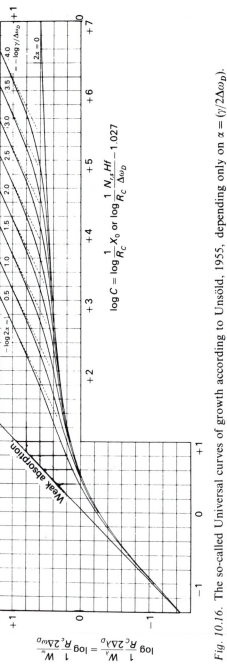

*Fig. 10.16.* The so-called Universal curves of growth according to Unsöld, 1955, depending only on $\alpha = (\gamma/2\Delta\omega_D)$. $R_c = R(\lambda_0)$ and $X_0 = X(\lambda_0)$.

This is the form of the curve of growth which Unsöld calls the 'Universal' curve of growth, which we show in Fig. 10.16.

### 10.9.5 *The influence of the damping constant on the curve of growth*

We still have to talk about the influence of the damping constant $\gamma$ on the damping part of the curve of growth. Clearly, the damping wings will be more pronounced if $\gamma$ is larger. Therefore, for larger $\gamma$ the damping part of the curve of growth will start for smaller values of $X(\lambda_0)$. The value of $\gamma$ is usually expressed by means of the parameter $\alpha$, with $\alpha = \gamma/2\Delta\omega_D$. The larger $\alpha$, the earlier the damping part will start. Fig. 10.16 shows the final curves of growth obtained for different values of $\alpha$, as given by Unsöld (1955).

# 11

# The hydrogen lines

The hydrogen lines deserve special consideration, not only because they are the strongest lines, but also because they are a very important tool in analyzing stellar spectra. The reason is that they are broadened by the so-called 'molecular' Stark effect, that is by the electric fields due to the passing ions. When a radiating atom or ion finds itself in an electric field, the energy levels are shifted to slightly different values of the excitation energy $\chi_n$ or $\chi_{exc}$. The emitted lines therefore occur at slightly different wavelengths. Even more important is that for the hydrogen atom the different orbitals contributing to the energy level with a main quantum number $n$, are normally degenerate, i.e., when there is no external force field they all have the same excitation energy (this is, of course, indicated by the statistical weight for each level). In the presence of an external force field this degeneracy is removed, so that in an external field the different orbitals contributing to a given level with main quantum number $n$ now occur at slightly different values for the excitation energy. This in turn means that, instead of one line being emitted in the force free case, there are now several lines emitted or absorbed at slightly different wavelengths. Fig. 11.1 shows the different components which are observed for the different Balmer lines.

This shifting of the energy levels in the electric field of the neighboring ions has mainly two effects. First, it *shortens the lifetime of the undisturbed level* which, according to equation (10.15), *results in a broadening of the undisturbed energy level*. Since this broadening is due to the electric field of the ions passing nearby or due to 'collisions' with other ions, this broadening is called *impact broadening*. Since this broadening is also due to a shortening of the lifetime of the original level, its effect on the width of the line is the same as that of radiation damping. It leads to a line profile which is similar to that for radiation damping but with a larger $\gamma$. Since the collision with the passing ion only takes a short time, the new, shifted levels only have a very short lifetime and are also broadened by this effect.

In addition to this impact broadening of the unshifted lines, we have to consider the effects of the new line components, for which the splitting increases with increasing field strengths. Ions passing very close cause a large splitting, while ions passing at larger distances only cause a small splitting. The pattern of the absorbed or emitted line therefore depends on the statistics of the electric fields which the absorbing or emitting atoms experience at the moment of the absorption or emission of the photon. In order to calculate the overall profile of the absorbed line, we have to look at the statistics of the electric fields experienced by the atoms at the moment

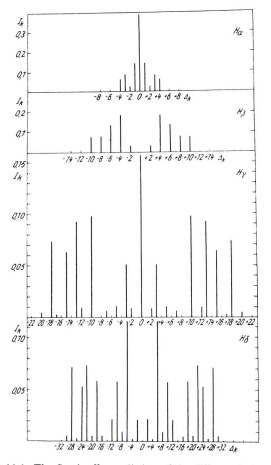

*Fig. 11.1.* The Stark effect splitting of the different Balmer lines (according to Unsöld, 1955, p. 320.)

of absorption. This part of the line broadening is described by the so-called *statistical line broadening theory*. Since it deals with the lines originating from the shifted energy levels, they occur in the wings of the original line. The larger the electrical fields or, in other words, the closer the encounters, the further out in the wing this displaced line occurs. The line profile in the wings of the hydrogen lines is therefore described by the statistical theory of the broadening by the 'molecular' Stark effect. It is clear that this broadening is not actually due to molecules, but rather to ions. This name is only used to indicate that it is not due to macroscopic electric fields, but rather to microscopic fields due to the charged particles in the plasma.

If we want to calculate the distribution of the absorption coefficient for the hydrogen lines, we have to calculate the probability for absorption at a given wavelength difference $\Delta\lambda$ from the line center. If this $\Delta\lambda$ is still within the impact broadened Lorentzian profile of the original line, then we know the absorption coefficient from the Lorentzian line profile; but if $\Delta\lambda$ is larger and is in the wings where the statistical broadening theory has to be applied, we have to determine the probability of finding an ion at the correct distance $d$ from the absorbing atom that creates a field $F = e/d^2$ leading to the shift $\Delta\lambda$ under consideration. We see immediately that the chance of finding a particle at the required distance $d$ is larger if there are more particles per $cm^3$. The chance of finding at a given moment an ion at exactly the required distance is proportional to the number of particles $N^+$ present per $cm^3$. We therefore find that in the case of the statistical line broadening in the hydrogen wings the absorption coefficient at any given $\Delta\lambda$ is proportional to the number of ions per $cm^3$ which equals the number of free electrons per $cm^3$, at least if all ions are ionized only once.

Why do we never mention the electric fields caused by the passing electrons? We said that the broadening is due to the shifting of energy levels in the atom due to the electric field experienced by the absorbing atom. In the case of the passing electrons the velocities of the electrons are too large. The collision times are too short. In other words, the phase shift $\eta = \int \Delta\omega\, dt$ experienced by a light wave is too small.

The actual profiles of the hydrogen line absorption coefficients have been calculated in successively better approximations. The latest ones were tabulated by Vidal, Cooper and Smith (1973). What is of interest to us here is the fact that at any $\Delta\lambda$ in the wings of the hydrogen lines $\kappa_\lambda$ is proportional to the number of free electrons per $cm^3$, $n_e$. For high density atmospheres when $n_e$ is large, the hydrogen lines are much broader than for low density atmospheres in supergiants (see Fig. 11.2).

Another important fact is the increase in the splitting of the different Stark effect components of the lines with increasing main quantum number $n$ (see Fig. 11.1). In fact, the splitting increases with $n^2$. For a given electric field the Balmer lines with a larger $n$ of the upper level are broadened more than H$\alpha$ or H$\beta$. On the other hand, we know that as the upper energy levels of the hydrogen atom get closer and closer together, the energy difference decreases as $1/n^3$. For some level $n = n_u$ the line broadening becomes larger than the difference in wavelength for successive Balmer lines. The lines merge together and can no longer be seen separately (see Fig. 11.2). From a simplified broadening theory for the Balmer lines – the original Holtsmark theory – Inglis and Teller (1939) find that the highest quantum number $n_u$ for which the energy levels are still separated is related to the electron density $n_e$ by

$$\log n_e = 23.26 - 7.5 \log n_u. \qquad 11.1$$

This equation provides a very simple means of determining the electron density $n_e$. We only have to count the number of visible Balmer lines, determine the upper quantum number of the last visible line $n_u$, and calculate $n_e$ from equation (11.1).

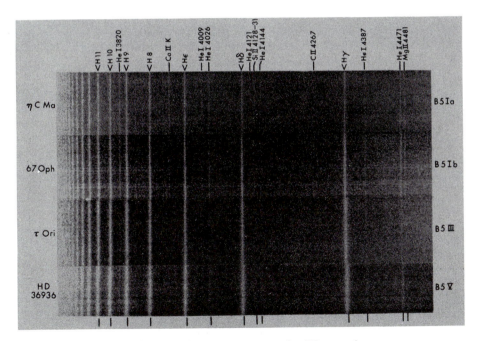

*Fig. 11.2.* The spectra of a B5 main sequence star and a B5 supergiant are compared. In the main sequence star the hydrogen lines are much wider than in the supergiant. For the supergiant we can see more Balmer lines separately. (From Morgan, Abt and Tapscott, 1978. This is a negative. White means no light.)

# 12

---

# Spectrum analysis

## 12.1  The Balmer jump and the hydrogen lines

### 12.1.1  Hot stars

In Section 8.2 we saw that the Balmer discontinuity is determined by the ratio of the continuous absorption coefficients on both sides of this discontinuity. For hot stars we derived that the discontinuity is determined by the ratio of the number of hydrogen atoms in the second quantum level (absorbing shortward of 3647 Å) to the number of hydrogen atoms in the third quantum level (absorbing on the long wavelength side of 3647 Å). This ratio is completely determined by the temperature. For the B stars the temperature can in principle be determined from the Balmer discontinuity alone.

The pressure can be determined from the electron density $n_e$. For late B stars, for instance, the hydrogen is completely ionized, while the helium is not. Therefore, we know that $n_e = H^+$ and $N = n_e + H^+ + He = n_e + H^+ + 0 \cdot 1 H^+$ if the abundance of helium is 10% by number of atoms. With this we find

$$N = 2.1 n_e, \qquad\qquad 12.1$$

and, of course,

$$P_g = NkT.$$

As we saw in the previous chapter, the number of free electrons can be determined from the hydrogen lines in two ways: either from the number of visible Balmer lines by means of the Inglis–Teller formula (equation 11.1), or by means of the hydrogen line wings. In Section 10.1 we saw that the line depth for optically thin lines is given by equation (10.8):

$$R_\lambda = \frac{2}{3} \frac{\kappa_L}{\kappa_c} \frac{d \ln B_\lambda}{d\tau_c}. \qquad\qquad 12.2$$

For the hot stars the continuous absorption coefficient $\kappa_c$ in the visual is due to the Paschen continuum, which means that it is due to absorption from the third level of the hydrogen atom. This means $\kappa_c$ is proportional to $N_H(n = 3)$. The line absorption coefficient for the Balmer lines which originate by absorption from the second quantum level is, of course, proportional to the number of hydrogen atoms in the second quantum level $N_H(n = 2)$. As we saw in the previous chapter, the line absorption coefficient in the wings of the Balmer lines is also proportional to the number of free electrons because of the Stark effect broadening of these lines. Therefore, we find for the hydrogen lines in the B stars

$$R_\lambda \propto \frac{N_H(n = 2)n_e}{N_H(n = 3)}.$$ 12.3

Knowing the temperature from the Balmer discontinuity, the depth in the Balmer line wings tells us the electron pressure and thereby the gas pressure.

Since the gas pressure must balance the weight of the overlying material, the gas pressure also gives a rough estimate for the gravitational acceleration in the stars. We must have approximately

$$P_g = mg,$$ 12.4

where $m$ is the mass of the overlying material.

Knowing that we are looking down into an optical depth $\tau_c = \frac{2}{3}$ and also knowing that the continuous absorption is due to the hydrogen atoms in the third quantum level, we can calculate the number of hydrogen atoms in the third quantum level above $\tau_c = \frac{2}{3}$ from

$$hN_H(n = 3) = \frac{2}{3}\frac{1}{\kappa_{ca}},$$ 12.5

where $\kappa_{ca}$ is the continuous absorption coefficient per hydrogen atom in the level $n = 3$, and $h$ is the effective height of the overlying atmosphere.

From the Boltzmann and Saha formula we can calculate the number of $H^+$ ions which constitute the main fraction of the mass. Considering the abundance of helium the total mass per $H^+$ ion is given by $m_H(1 + 0.1 \times 4) = 1.4m_H$ and the total mass $m$ above the level $\tau_c = \frac{2}{3}$ is given by

$$m = h \times 1.4 \times N_{H^+} \times m_H.$$ 12.6

An approximate value for the gravity $g$ is then obtained from

$$g = P_g/m.$$ 12.7

In determining the mass per $H^+$ ion we have already made use of the abundance of the elements which we only know after we have made a complete analysis. We will discuss this shortly.

In Section 8.2 we discussed the dependence of the Balmer discontinuity on pressure and temperature. We saw that for B stars the size of the discontinuity is determined by

$$\frac{\kappa(3647^+)}{\kappa(3647^-)} = \frac{\kappa^+(\mathrm{H})N_\mathrm{H}(n=3)}{\kappa^-(\mathrm{H})N_\mathrm{H}(n=2)} \propto \frac{N_\mathrm{H}(n=3)}{N_\mathrm{H}(n=2)} = \frac{g_3}{g_2}\mathrm{e}^{-(\chi_3-\chi_2)/kT},$$

which means the discontinuity depends *only* on the temperature, which can therefore be determined from the Balmer jump alone. However, for O stars and for supergiants the importance of the electron scattering introduces some dependence on the electron density.

### 12.1.2   *The cool stars*

In the cool stars the continuous absorption coefficient is determined by the negative hydrogen ion $\mathrm{H}^-$. We discussed in Section 8.2 that for these stars the Balmer discontinuity is dependent on both the temperature and the electron pressure, because the absorption on the long wavelength side of the Balmer jump is due to $\mathrm{H}^-$ and the number of $\mathrm{H}^-$ ions increases with increasing electron density as given by the Saha equation.

Let us now look at the depth of the Balmer line wings in these stars. We again start from equation (12.2). The continuous absorption is now due to $\mathrm{H}^-$, and the line absorption coefficient in the Balmer line wings is again proportional to the number of hydrogen atoms in the second quantum level, and to $n_e$, because of the Stark effect broadening. We thus have

$$R_\lambda \propto \frac{N_\mathrm{H}(n=2)n_e}{N(\mathrm{H}^-)} \propto \frac{N_\mathrm{H}(n=2)n_e}{N_\mathrm{H}(n=1)n_e\varphi(T)} = \frac{N_\mathrm{H}(n=2)}{N_\mathrm{H}(n=1)\varphi(T)}. \qquad 12.8$$

Here we have made use of the Saha equation, which tells us that the number of $\mathrm{H}^-$ ions is proportional to $N_\mathrm{H}(n=1)n_e$ times a function $\varphi$ of $T$. For cool stars the depth in the Balmer line wings is only a function of temperature. We have for cool stars the inverse situation from what is found in hot stars. Now the Balmer discontinuity depends on $T$ and $n_e$ and the Balmer line wings depend on temperature alone. This is shown by the spectra in Fig. 12.1.

Of course, these considerations apply only to normal stars. For very luminous stars the Balmer lines sometimes show emission cores. For such stars we see immediately that our simple 'normal' atmosphere theory does not apply. Also, we have not talked about O stars, for which electron scattering becomes very important and for which our assumption that the source function equals the Planck function needs checking. It appears that this assumption is not true for these hot stars. A basic discussion of NLTE effects, including the derivation of the source function in general, will be given in Chapter 14.

For normal stars we have seen in this section that the Balmer jump and the hydrogen line wings together always determine the electron density and the temperature. We do not need to know anything else. What we determine are, of course, some average values for the atmosphere. In Fig. 12.2 we summarize the situation.

## 12.2 The Strömgren colors

In order to make use of these fortunate circumstances, Strömgren (1963) developed a color system which is especially designed to measure the Hβ line wings and the Balmer jump. In order to determine the

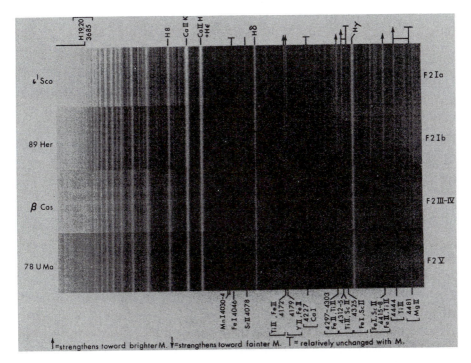

*Fig. 12.1.* For F stars the widths of the hydrogen lines are independent of the luminosity of the stars. (From Morgan, Abt and Tapscott, 1978.)

|  | Balmer discontinuity $D$ | Hydrogen line wings $H_\beta, H_\lambda$, etc. |
|---|---|---|
| Hot stars | $D = D(T)$ | $R_\lambda = R_\lambda(n_e, T)$ |
| Cool stars | $D = D(n_e, T)$ | $R_\lambda = R_\lambda(T)$ |

*Fig. 12.2* summarizes the dependences of the Balmer jump and the depths of the hydrogen line wings on temperature and electron density for hot and for cool stars.

Table 12.1. *Central wavelengths and bandwidths of intermediate bandwidth photometry used by Strömgren*

| Band | $\lambda$ (Å) | Halfwidth (Å) | Band | $\lambda$ (Å) | Halfwidth (Å) |
|------|---------------|---------------|------|---------------|---------------|
| u | 3500 | 300 | b | 4670 | 180 |
| v | 4110 | 190 | y | 5470 | 230 |

Table 12.2. *Central wavelengths and bandwidths used in the Strömgren narrow-band photometry*

| Filter | | $\lambda_{max}$ (Å) | Halfwidth (Å) |
|--------|--------------------|---------------------|---------------|
| a | Interference filter | 5030 | 90 |
| b | Interference filter | 4861 | 35 |
| c | Interference filter | 4700 | 100 |
| d | Interference filter | 4520 | 90 |
| e | Interference filter | 4030 | 90 |
| f | Glass filter | 3650 | 350 |

abundances of the heavy elements, he also included a blue band which is particularly sensitive to the metallic line absorption. The wavelength bands used in the Strömgren photometry are much narrower than UBV color wavelength bands. With Strömgren photometry not quite such faint stars can be reached as with the UBV photometry, but better physical information can be obtained. Strömgren actually uses two color systems: one with intermediate bandwidths ($\Delta\lambda \approx 200$ Å) and one with still narrower bands ($\Delta\lambda \approx 100$ Å). The names, central wavelengths and bandwidths for both systems are given in Tables 12.1 and 12.2.

The Balmer jump (see Fig. 12.3) is found in both systems by measuring the intensity gradient on the long wavelength side, which would be about the expected gradient across the Balmer jump if there were no discontinuity, and comparing that with the actually measured gradient across the Balmer jump. In the intermediate band system the Balmer jump index is defined as

$$c_1 = (u - v) \qquad\qquad\qquad - (v - b)$$

$$\text{actual gradient across}\qquad \text{long wavelength}$$
$$\text{the Balmer discontinuity}\qquad \text{gradient}$$

where the letters stand for the magnitudes measured with the corresponding filters.

In the narrow band system the Balmer jump index is defined as

$$c = (f - e) \qquad\qquad - (e - d)$$
$$\text{gradient across the} \qquad \text{long wavelength}$$
$$\text{Balmer discontinuity} \qquad \text{gradient}$$

The strength of the Balmer lines, which is determined by the strengths of the wings, is measured only in the narrow band system. Strömgren defines the $l$ index as

$$l = b - \tfrac{1}{2}(a + c),$$

which means he measures the flux in the H$\beta$ band with H$\beta$ ($\lambda = 4861$ Å) in the center and compares it with the average intensity on both sides (see Fig. 12.4).

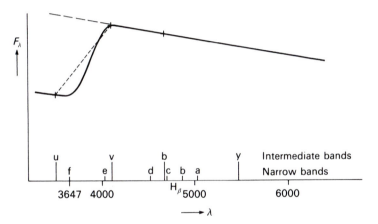

Fig. 12.3. The Strömgren bands describe the gradient of the energy distribution on both sides of the Balmer jump. The comparison of the extrapolation from the long wavelength side with the actual gradient over the discontinuity measures the Balmer jump.

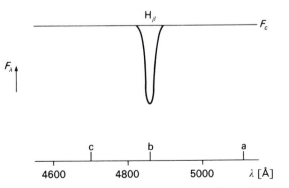

Fig. 12.4. The H$\beta$ index compares the average flux and both sides of H$\beta$ with the flux in H$\beta$.

Crawford (1958) modified this method and measured the flux through two filters, both centered at H$\beta$, but one with a bandwidth of 15 Å containing only intensities within the broad H$\beta$ line, and one filter with a bandwidth of 150 Å, in which most of the light comes from wavelengths on both sides of H$\beta$. He defines an H$\beta$ index as

$$\beta = (m(15 \text{ Å}) - m(150 \text{ Å})).$$

In the literature one usually finds the use of $\beta$ and $c_1$, indicating a combination of both systems.

Our previous discussion of the Balmer lines tells us that these two indices determine $T$ and $n_e$ for all 'normal' stars.

The Strömgren color system is able to measure minor differences in the Balmer discontinuity. For cool stars rather minor differences in the electron density can therefore be measured, which is especially useful for the determination of stellar ages, as we shall discuss in Volume 3.

In the intermediate bandwidth system the b − y colors can serve to replace the B − V colors and can also be used to determine the temperature of the stars.

The overall strengths of the metallic lines, which can be used as a measure of the overall metal abundances, can be measured in the intermediate bandwidths system by the $m_1$ index defined as

$$m_1 = (v - b) - (b - y).$$

This index makes use of the fact that the line intensities are stronger in the blue and ultraviolet spectral region than in the yellow and red. It compares the intensity gradient at the shorter wavelengths, where it is reduced by the absorption lines, with the intensity gradient at longer wavelengths, where it is influenced very little by the lines.

In the narrow band photometry a similar index $m$ can be defined by

$$m = (e - d) - (d - a),$$

which makes use of the same facts.

There are some problems with a simple interpretation of the $m$ and $m_1$ indices. As we saw in the previous chapter, the line strength on the flat part of the curve of growth may be increased due to microturbulence. It is not quite clear how much the $m$ index might be influenced by this effect. Stars with large microturbulence or with magnetic fields could possibly be mistaken for stars with a very large abundance of heavy elements. All color systems designed to measure heavy element abundances from the integral effects of many strong lines suffer from these limitations.

## 12.3    The curve of growth analysis

### 12.3.1  *Line identification*

In order to determine abundances, we have to identify the lines, i.e., we have to know to which elements they belong and to which transition they correspond. Therefore, first we have to measure the wavelengths of the lines under consideration. This is done by measuring the positions of lines of 'laboratory' spectra, also called 'comparison' spectra, which are photographed together with the stellar spectrum. These laboratory spectra are actually generated in a light source right at the telescope, such that the spectra can be observed through the same spectrograph and photographed and developed on the same photographic plate as the stellar spectrum (see Fig. 12.5). The wavelengths of the spectral lines in the laboratory spectra are predetermined. The positions of the stellar spectral lines then have to be measured accurately. The wavelengths of the stellar lines can then be determined by interpolation between the known wavelengths of the laboratory spectral lines.

Once we know the wavelengths, we have to identify the lines. There are tables that give all the wavelengths of the known lines and also give the corresponding transitions. Such a table is Miss Moore's *Multiplet Tables* (1959). At the back of these tables, we find a list of lines ordered according to wavelength. Once we have the measured wavelength of the line under consideration to within a certain uncertainty, which is determined by the dispersion and the general quality of the spectrum (the uncertainty is usually of the order of 0.1 Å if you have a high resolution spectrum with about 4 Å per mm), there are usually about three to four lines in that given uncertainty interval of about 0.1 Å between which we have to choose. For

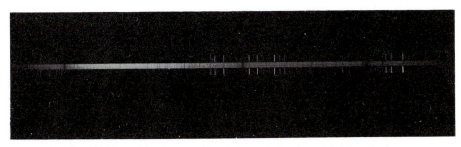

*Fig. 12.5.* Photograph of the stellar spectrum and of 'laboratory' spectra of an iron arc on both sides. The wavelengths for the spectral lines in the 'laboratory' spectrum have been determined previously. The wavelengths of the stellar lines can be determined by accurate interpolation.

each line in the table the line identification is given according to the element and to transition. This means the number of the multiplet is given to which this line belongs. In the multiplet tables one can then look up what the excitation energies are for the two levels involved in the transition and the statistical weights for the levels can also be obtained.

What is a multiplet?

In the hydrogen atom with only one electron all the orbitals with the same main quantum number $n$ have the same energy. The energy levels are degenerate. In more complicated atoms with many electrons this is not the case. For the same electron configuration we can find different values for the total angular momentum $J$, depending on how the angular momenta of the different electrons are oriented with respect to each other. For a given orbital angular momentum $L$ and a given spin $S$ we still obtain different energy levels, depending on the value of the total angular momentum $J$. Each energy level with given values of $L$ and $S$ still splits up into several fine structure levels according to the orientation of $L$ and $S$. Now consider two levels with different values of $L$ and $S$. For each level we have the fine structure splitting according to the $J$ values. All the lines originating from transitions between these fine structure levels of the two main levels are called one multiplet (see Fig. 12.6). For the lines in one multiplet the excitation energies of the different fine structure levels are

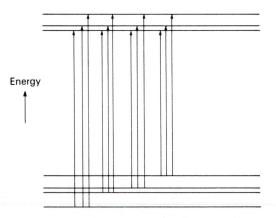

Energy

*Fig. 12.6.* For given values of the total orbital momentum $L$ and the spin $S$ the atomic energy level splits up into several fine structure levels according to the different values of $J$ which depends on the relative orientation of $S$ with respect to $L$. Lines due to transitions between the fine structure levels of two main levels are called a multiplet. All the transitions belonging to one multiplet are shown. Many of these transitions are 'forbidden', however, which means the transition probabilities are extremely small.

nearly equal. The relative line strengths for the lines within each multiplet are therefore completely determined by the transition probabilities or by their relative oscillator strengths. The relative intensities are therefore independent of temperature and pressure. This fact is very important for the spectrum analysis by means of the curve of growth.

Let us go back to the problem of line identification. Suppose the measured wavelength of a line at 3943.28 Å, which you want to identify, almost agrees with the wavelength of an iron line which belongs to the multiplet with the number 72. In order to check the identification you have to go to the front of the multiplet tables and look up the multiplet number 72 of iron. You find there are three lines in this multiplet which are stronger than the line you have measured. If your identification of the line under consideration with the iron line is correct, you should also see these three stronger lines in your spectrum at $\lambda 3949.9$, 3977.74 and 4003.71 Å. They must be stronger than the line you have measured. You have to find the wavelengths of those three lines in your spectrum and see whether these lines are there and whether they are indeed stronger than the line you have measured. If they are, then the identification is possibly correct. If the three lines are not there or are too weak, then the identification of the measured line as this iron line is most probably wrong and you should check whether the line next to it in the multiplet table – namely, the Sm II line at 3943.239 – might be a better possibility. Only if all the relative line strengths within the multiplets in question match the expectations can you be relatively certain that you have made the correct identification.

### 12.3.2 *Fitting the lines of one multiplet on the curve of growth*

In principle the curve of growth analysis consists of measuring the equivalent width, dividing by $\Delta\lambda_D$ and by $R_c$, the maximum central depth of the strongest lines, and determining the ordinate in the curve of growth for a given line. We can then fit this value on the theoretical curve of growth shown in Fig. 10.16. The point where it fits on the curve of growth determines the abscissa. Because the abscissa depends on the number of absorbing atoms the value obtained for the abscissa tells us what the number of absorbing atoms is. The only problem in this procedure is that we do not know *a priori* what the Doppler width $\Delta\lambda_D$ is. Depending on the assumption about $\Delta\lambda_D$, we may fit a point for a strong line on a very different part of the curve of growth, as shown in Fig. 12.7. We therefore have to determine the value for $\Delta\lambda_D$ very carefully. Only for very weak lines on the linear part of the curve of growth do we find the same number of absorbing atoms for a given equivalent width independently of the assumed value for the Doppler widths, because abscissa and ordinate are divided by the same factor, but such weak lines are hard to measure.

In addition, we would also like to determine the degree of excitation in the atoms and ions, in order to determine the total number of particles of a given kind and not just the number of particles absorbing in a given line.

In the following discussion we want to show how the Doppler width can also be determined by means of the curve of growth and to show how the degree of excitation can be determined. We start by looking at the lines of one multiplet only.

Once you know the correct line identification you have to look up the oscillator strengths of the lines in another table. Oscillator strengths can either be calculated by means of quantum mechanics which in the case of complicated atoms is difficult and not very accurate. For the hydrogen atom theoretical oscillator strengths are very well determined. For the complicated atoms like iron measurements appear to be more accurate, though the laboratory light sources usually show temperature variations within the light source, which makes it difficult to determine the degree of excitation which determines the number of emitting atoms. If this number was determined inaccurately then the measured oscillator strengths are all wrong. Because of this problem the first measured iron oscillator strengths for iron lines with higher excitation potentials were all off by nearly a factor of ten and correspondingly the measured iron abundances were all wrong by the same factor. We only say this to show that the measurement of

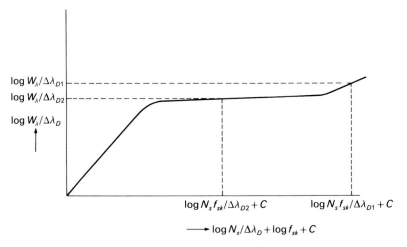

*Fig. 12.7* shows a schematic curve of growth. Different values of $\Delta\lambda_D$ may lead to very different values for the derived number $N_s$ of absorbing atoms.

good oscillator strengths is not an easy task and we are still lacking good values of oscillator strengths for many lines of complicated atoms. But once we have reasonably good values for the oscillator strengths, we can fit the lines of a given multiplet on the theoretical curve of growth shown in Fig. 10.16. The abscissa of this curve of growth is the line absorption coefficient per cm$^3$ in the line center times a factor $\frac{2}{3} \times 1/\kappa_c \times d \ln B_\lambda/d\tau_c$ and then the normalizing factors $1/\Delta\lambda_D$ and $1/R_c$, where $R_c$ is the largest possible central depth of a line. If we consider only a narrow spectral region, then all the last factors mentioned are nearly the same for the different lines. For a small wavelength band we only have to consider the different values for the line absorption coefficient when comparing different lines. The line absorption coefficient in the line center is given by a constant $\kappa_0$ and the number of absorbing atoms times the oscillator strength, i.e.,

$$\kappa_L = \kappa_0 N(n = n_i) f_{ik}, \qquad 12.9$$

where $n_i$ stands for the quantum numbers of the lower level involved in the transition, and $f_{ik}$ is the oscillator strength for the transition from the level with quantum numbers $i$ to the level with quantum numbers $k$. If we plot the abscissa in a logarithmic scale, then the difference in the abscissa for different lines is given by

$$\log N(n = n_s) - \log N(n = n_l) + \log f_{sm} - \log f_{lj}$$

if the line transitions are from $s$ to $m$ and from $l$ to $j$. The number of atoms in the various states of excitation can be described by the Boltzmann formula, namely

$$\frac{N(n = n_s)}{N_0} = \frac{g_s}{g_0} e^{-\chi_s/kT}, \qquad 12.10$$

where $g_s$ is the statistical weight for the level with quantum numbers $s$, $\chi_s$ is the excitation energy for this level, $N_0$ is the number of atoms in the ground level, and $g_0$ is the statistical weight of the ground level. For lines of a given multiplet the $\chi_s$ are all the same for different lines because the fine structure splitting of the levels is very small. With this in mind we can then say that the difference in the abscissa for lines of a given multiplet is given by

$$\Delta \text{ abscissa} = \log g_s + \log f_{sm} - \log g_l - \log f_{lj}, \qquad 12.11$$

or

$$\Delta \text{ abscissa} = \log g_s f_{sm} - \log g_l f_{lj}, \qquad 12.12$$

because $N_0$ and $g_0$ are also the same for one multiplet. In fact, they are the same for all lines of a given atom or ion.

It is always the product $gf$ which we need and it is also this product which is measured from the line intensities in the laboratory. Therefore, this product is usually given in the oscillator strength tables.

For the lines in a given multiplet we then know the difference in the abscissa. The equivalent widths have to be measured on the spectrum. To begin with, we can plot the first line $sm$ (going from quantum state $s$ to the state $m$) of the multiplet at an arbitrary position of the abscissa in the plot with the curve of growth, because we do not know the exact value for the abscissa. When we now plot the value for the second line $lj$ we know that its abscissa must be decreased by the amount $\log g_s f_{sm} - \log g_l f_{lj}$, as compared to the abscissa for the first line. We therefore know the relative position of this line with respect to the first line. For the third and fourth lines of the same multiplet we also know the relative values of the abscissa. After we have plotted all the values for the different lines of the multiplet we have a picture, as shown in Fig. 12.8 by the $\times$. Clearly, these points must fit on the theoretical relation between the equivalent widths and the abscissa. We see that we have plotted all our points for the wrong absolute value of the abscissa, which was to be expected, since we chose the abscissa of the first line arbitrarily. Shifting the points over onto the theoretical curve of growth without changing the relative values of the abscissa tells us where these points really belong on the curve of growth.

### 12.3.3  Determination of $\Delta\lambda_D$

We have still not talked about how to determine the ordinate which is not solely determined by the equivalent width – which we can

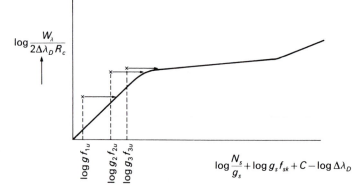

Fig. 12.8. For three lines of one multiplet the values of $\log gf$ are known and therefore their distances on the abscissa are known. The shift over onto the theoretical curve of growth determines the absolute values for the abscissa of the three lines.

measure – but also by the Doppler width $\Delta\lambda_D$ and the largest central depth $R_c$. On high resolution spectra for slowly rotating stars $R_c$ can be measured directly on the spectrum for the strongest lines in the spectral region being studied. (Remember that $R_c$ is wavelength dependent.) For low resolution spectra and for spectra of rapidly rotating stars this is not possible. We should then take the values measured for similar stars which are slow rotators and for which high resolution spectra are available. The intrinsic $R_c$ are rather similar for different stars even of slightly different spectral types.

For a determination of $\Delta\lambda_D$ we started assuming that we see only thermal Doppler broadening, which means $\Delta\lambda_D = \lambda\xi_{th}/c$, where $c$ is the velocity of light. If this Doppler broadening is too small, i.e., if there is microturbulence broadening in addition, then the equivalent widths for the lines on the flat part of the curve of growth are larger than corresponding to the thermal $\Delta\lambda_D$. They then appear too large in comparison with the equivalent widths of the weaker lines (see Fig. 12.9). If in a given multiplet we have weak lines *and* strong lines such that the horizontal shift is determined by the weak lines, the error in $\Delta\lambda_D$ can be obtained from the strong lines (as shown in Fig. 12.9). Once the $\Delta\lambda_D$ is determined from one multiplet, the same value can be used for the other multiplets of the same atom or ion in the same spectral region. The absolute value of the abscissa determines the number of atoms $N(n = n_s)$ absorbing in the lines of this multiplet. This number can be read off, except for a factor $C$, determined by the continuous $\kappa$ and the gradient of the Planck function. This factor $C$ is, however, the same for all lines in a given small wavelength region.

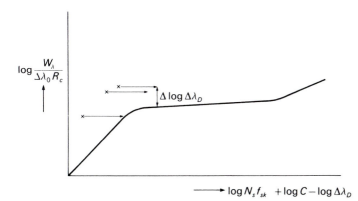

*Fig. 12.9.* If the assumed value for $\Delta\lambda_D$ was too small the points for the saturated lines lie too high by the error in $\Delta \log \Delta\lambda_D$, when the absolute value for the abscissa is determined by the weak lines.

*12.3.4  Determination of the excitation temperature $T_{exc}$*

After we have fitted one multiplet onto the curve of growth, we can proceed to do the same with the next multiplet and determine the number of atoms $N_n$ absorbing in this multiplet with an excitation energy $\chi_n$, except again for the factor $C$. The ratio of the occupation numbers for the levels $s$ and $n$ can then be determined because the common factor $C$ cancels out. Inserting this ratio into the Boltzmann formula, we find

$$\frac{N_n}{N_s} = \frac{g_n}{g_s} e^{-(\chi_n - \chi_s)/kT_{exc}}. \qquad 12.13$$

The temperature in the exponent is again some average value of the temperature in the atmosphere. This average is taken with different weighting factors than the temperature, which we determined from the Balmer discontinuity and the Balmer lines. Therefore, it does not necessarily have to be exactly the same as the Balmer line temperature, although the differences should never be large; at least not if local thermodynamic equilibrium is a good approximation. In order to make clear how this temperature was determined, we call this temperature the *excitation temperature*, because it is determined from the excitation of the different energy levels. Even if LTE should not be a good approximation, excitation temperatures can always be defined by means of equation (12.13). But the values obtained with occupation numbers for different levels can then be quite different. In fact, the excitation temperatures obtained for different energy levels can serve as one test, whether the assumption of local thermodynamic equilibrium is valid. If the excitation temperatures disagree distinctly, then the assumption of LTE is not a good one.

*12.3.5  Determination of the ionization temperature*

When we have determined the occupation number $N_s$ of a given level and also the excitation temperature $T_{exc}$, we are in the position of being able to calculate the occupation numbers for all other levels for that atom or ion for which we cannot observe lines. With this we calculate the total number of atoms of the kind for which we have studied the lines. Again, this number still has the unknown factor $C$ attached. Suppose we have made this study for the neutral iron lines. We can then make the same kind of investigation for the lines of the ionized iron and determine the total number of $Fe^+$ ions. If we know what the electron pressure is – for example, from the Balmer discontinuity or from the Inglis–Teller formula – we can use the Saha equation, plug in the ratio of $Fe^+/Fe$ and determine the temperature from the Saha equation. Because this temperature is determined from the degree of ionization, it is called the *ionization temperature*. Again, this is an average over the photosphere with slightly different weights in the averaging procedure. This temperature may be slightly different from the excitation temperature. We can go one step further and do the same procedure for another element, for instance, titanium. For titanium the $\Delta\lambda_D$ is probably different from the value obtained for iron because its atomic weight is different. Therefore, the thermal velocity is different. We have to determine a new $\Delta\lambda_D$ for titanium. So we again find a multiplet with many lines of different equivalent widths such that we can determine a curve of growth from one multiplet. If we do not find a suitable one we can actually combine several multiplets. We know that the occupation numbers of the lower levels are determined by the Boltzmann formula. If the first multiplet originates from a level with quantum number $n$ and excitation energy $\chi_n$, and the other multiplet originates from a level with quantum number $m$ and excitation energy $\chi_m$, then we know from the Boltzmann formula that the ratio of the occupation numbers $N_n$ and $N_m$ is given by

$$\log N_m - \log N_n = \log g_m - \log g_n - \Theta(\chi_m - \chi_n), \qquad 12.14$$

or

$$\log \frac{N_m}{g_m} - \log \frac{N_n}{g_n} = -\Theta(\chi_m - \chi_n). \qquad 12.15$$

If we use the number of absorbing atoms in the level $n$, namely $N_n$, in the abscissa of the curve of growth for both multiplets, then we have to express the number of atoms $N_m$ absorbing in the second multiplet by means of $N_n$ and the Boltzmann formula, as given by equation (12.15), which gives the difference in abscissa for the two multiplets. Unfortunately, there is the quantity $\Theta = 5040/T_{\text{exc}}$ in equation (12.15), and we do not yet know its value. We can, however, try different values of $\Theta$ and see for which $\Theta$ the two multiplets fit best on one curve. If the assumption of LTE is valid, then the same $\Theta$ should also hold for other multiplets. We can therefore use as many multiplets as we want and determine $\Theta$ from the condition that all the multiplets should fit on one curve of growth with the smallest amount of scatter. This can also be done graphically, as demonstrated in Fig. 12.10. As previously shown, we plot $\log(W_\lambda/(\Delta\lambda_D R_c)$ as a function of $\log(gf)_{ni}$ for the different lines with index $i$ of the multiplet from level $n$. We then do the same for the lines with index $j$ from the level $m$. Since $\log N_m/g_m$ and $\log N_n/g_n$ differ by $\Theta(\chi_m - \chi_n)$ as given by equation (12.15), the abscissae for the two multiplets differ by this amount, which means the two curves of growth outlined by the two multiplets are shifted with respect to each other by this amount. The horizontal shift necessary to fit one multiplet on top of the other determines $\Theta(\chi_m - \chi_n)$. With the known excitation energies the value of $\Theta$ can be obtained.

The excitation temperature can thus be obtained by fitting the different multiplets together.

### 12.3.6  *The determination of the kinetic temperature*

In Section 12.3.3 we saw that $\Delta\lambda_D$ – the Doppler width – given by the thermal motions and by the microturbulent motions, can be determined from the vertical shift necessary to fit the observed curve of growth onto the theoretical curve of growth. We also emphasized that the value for $\Delta\lambda_D$ must be expected to be different for different elements with different atomic weights, because their thermal velocities are different. In principle, this difference in thermal motion for different particles can be used to determine the kinetic temperature in the gas, which we would generally consider to be *the* temperature of the gas.

We saw in Section 10.1.5 that $\Delta\lambda_D$ is given by

$$\Delta\lambda_D/\lambda = \xi/c \qquad \text{with } \xi^2 = \xi_{th}^2 + \xi_{turb}^2, \qquad\qquad 12.16$$

where $\xi_{turb}$ is the microturbulent reference velocity and where the thermal reference velocity $\xi_{th}$ is given by

$$\xi_{th}^2 = \frac{2R_g T}{\mu}, \qquad\qquad 12.17$$

$\mu =$ atomic weight.

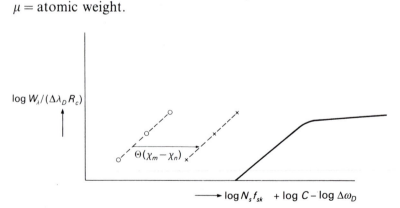

*Fig. 12.10.* The equivalent width for the lines in multiplet $n$ and $m$ divided by $\Delta\lambda_D R_c$ are plotted as a function of $\log g_{ni}f_{ni}$ and $\log g_{mj}f_{mj}$, respectively. Their abscissae in the curve of growth differ by $\log N_m/g_m - \log N_n/g_n = -\Theta(\chi_m - \chi_n)$ which can be determined by the horizontal shift necessary to shift the points for the two multiplets on top of each other.

For particles with large $\mu$ the thermal velocity is small, while particles with a small molecular weight $\mu$ have large velocities. It is this difference which, in principle, makes it possible to determine the kinetic temperature. Suppose we have measured $\Delta\lambda_D$ for iron lines and for carbon lines. We then know $\xi^2(\text{Fe})$ and $\xi^2(\text{C})$, where

$$\xi^2(\text{Fe}) = \frac{2R_g T}{\mu(\text{Fe})} + \xi^2_{\text{turb}},$$ 12.18

and

$$\xi^2(\text{C}) = \frac{2R_g T}{\mu(\text{C})} + \xi^2_{\text{turb}}.$$ 12.19

From equations (12.18) and (12.19) we then find

$$T = \left(\frac{\xi^2(\text{Fe})}{2R_g} - \frac{\xi^2(\text{C})}{2R_g}\right)\left(\frac{1}{\mu(\text{Fe})} - \frac{1}{\mu(\text{C})}\right)^{-1}.$$ 12.20

In practice, there are limits to the accuracy of this temperature determination. First of all, this can only work if the microturbulent velocities $\xi_{\text{turb}}$ are not much larger than the thermal velocities. Otherwise, the differences in $\Delta\lambda_D$ are too small to be measurable. The second problem is that the atomic data, as well as the measured equivalent widths, are often not very accurately known, which means that the best vertical shift between observed and theoretical curve of growth is often not very accurately determined, resulting in rather large uncertainties for $\Delta\lambda_D$ and therefore for $\xi^2$. However, in the future improved atomic data and higher accuracy of the measurements will make this determination of a kinetic temperature a nice tool to check on our assumption of LTE.

### 12.3.7 The determination of the electron pressure from the curve of growth

In Section 12.3.5 we discussed how we can determine the ionization temperature of, for instance, iron, if from the curve of growth analysis we have determined both the number of neutral iron atoms Fe and the number of ionized iron ions $Fe^+$. We stated that, having obtained the number of free electrons $n_e$ from a study of the Balmer lines and/or the Balmer discontinuity, we can use the Saha equation to calculate the ionization temperature $T_{\text{ion}}$. Here we show that the electron density $n_e$ can, in principle, also be determined from the curve of growth analysis. Suppose we have not only determined the number of iron atoms and ions, but also the number of titanium atoms and ions. We can then use the Saha equations for both elements, which gives us two equations to determine both unknowns: the ionization temperature and the electron density. The division of both Saha equations gives one equation for the determination of the ionization temperature:

$$\log(\mathrm{Fe}^+/\mathrm{Fe}) - \log(\mathrm{Ti}^+/\mathrm{Ti})$$
$$= \log(u^+/u)_{\mathrm{Fe}} - \log(u^+/u)_{\mathrm{Ti}} - \Theta(\chi_{\mathrm{ion}}(\mathrm{Fe}) - \chi_{\mathrm{ion}}(\mathrm{Ti})). \quad 12.21$$

Once the ionization temperature, or $\Theta$, has been determined from this equation, the electron density can be obtained from either of the two Saha equations.

As is obvious from equation (12.21), the determination of $\Theta$, and therefore the ionization temperature, by this method is very inaccurate if the ionization energies of the elements being used are nearly equal. Small errors in the empirical determination of the left-hand side of the equation lead then to very large errors in the value for $\Theta$ and $T_{\mathrm{ion}}$. When using this method we should therefore always try to use element pairs with very different ionization energies. Unfortunately, this is not always possible, because for elements with very low ionization energies the atomic lines may no longer be visible, while for those with high ionization energies the ionic lines are too weak to be measurable. The elements or ions observable in both stages of ionization usually have ionization energies which do not differ very much. Therefore, it is often preferable to determine the electron density from the Balmer lines and discontinuity and determine the ionization temperature from a single Saha equation.

### 12.3.8    Dependence of some line strengths on temperature, gravity and element abundances

In Volume 1 we pointed out that the temperature of a star has a large influence on the strengths of the different spectral lines. For high temperature stars there are no lines of neutral elements in the spectrum because the atoms are all ionized except for helium, which has a very high ionization energy, and except for hydrogen, which is so very abundant and which also determines the continuous absorption coefficient. The line depth $R_\lambda$ depends on the ratio of the line absorption coefficient to the continuous absorption coefficient, which for the Balmer lines means

$$R_\lambda \propto \frac{\kappa_L}{\kappa_c} \propto \frac{N_\mathrm{H}(n=2)}{N_\mathrm{H}(n=3)} n_e = \frac{g_2}{g_3} \mathrm{e}^{-(\chi_2 - \chi_3)/kT} n_e = \frac{g_2}{g_3} \mathrm{e}^{(\chi_3 - \chi_2)/kT} n_e, \quad 12.22$$

if the continuous absorption is due to the neutral hydrogen atoms in the third quantum level, as, for example, in early B stars. For increasing $T$ the $R_\lambda$ decreases. For increasing degree of ionization of hydrogen, both the line absorption and the continuous absorption coefficients decrease and the ratio only changes because the occupation number in the third quantum level, doing the continuous absorption, increases relative to the one of the second quantum level responsible for the line absorption. This is the reason for the decrease of the Balmer line intensities with increasing temperature for the B stars. The temperature dependence is very weak because $\chi_3 - \chi_2$ is so small.

When comparing line strengths in supergiants with those of dwarfs, we have to be aware of the influence of the gravity which determines the pressure in the atmospheres. For instance, for G type stars the continuous absorption is due to the $H^-$ ion. The Saha equation tells us that the number of $H^-$ ions relative to the number of neutral atoms increases with increasing electron density $n_e$. Supergiants therefore generally show stronger spectral lines than dwarf stars (see Fig. 12.1), not because there are more heavy elements in their atmospheres, but because the continuous absorption coefficient is smaller in these atmospheres of low gravity and therefore small electron density. Therefore, we have to be very cautious when looking at stellar spectra. Weak lines do not necessarily mean a low abundance of heavy elements and vice versa.

For the Balmer lines of hydrogen we find, for instance, for A and B stars that the lines become *stronger* when the *hydrogen abundance decreases* if the hydrogen is replaced by helium. The reason is the increase in gas pressure coupled with an increase in the electron pressure in the atmospheres with larger helium abundances. Because the line absorption coefficient in the Balmer line wings is proportional to the electron density and the continuous absorption coefficient is still due to the hydrogen, we find an increase of the hydrogen line strength for stars with a smaller hydrogen abundance, unless we go to extremely small hydrogen abundances, when the continuous $\kappa$ becomes due to helium.

When making a computer analysis of stellar spectra, we have to be aware that when trying different abundances of heavy elements in order to match the observed spectrum we have to be sure to change the atmospheric model each time we try a different element abundance because the change in the continuous absorption coefficient due to the changing metal abundance is as important as the change in the line absorption coefficient.

As an example, let us look at the dependence of an $Fe^+$ line strength on the abundance of Fe for solar type stars. For such a line the line depth is determined by the ratio

$$R_\lambda(Fe^+) \propto \frac{Fe^+}{H^-} \propto \frac{Fe^+}{N_H n_e}. \qquad 12.23$$

In solar type stars the electrons come from the ionization of the metals like iron. We saw that in the sun iron is mainly ionized. The iron abundance is therefore given by

$$\text{abundance (Fe)} = Fe^+/N_H = A(\text{Fe}).\qquad\qquad 12.24$$

If the abundances of all the heavy elements vary together, as is roughly the case, then we must also have

$$n_e \propto Fe^+ = A(\text{Fe})N_H.\qquad\qquad 12.25$$

Inserting this into equation (12.23), we find

$$R_\lambda(Fe^+) \propto \frac{A(\text{Fe})N_H}{N_H n_e} = \frac{A(\text{Fe})}{n_e} \propto \frac{1}{N_H}.\qquad\qquad 12.26$$

The Fe abundance cancels out in a first approximation. Actually, $N_H$ increases for decreasing metal abundances. Therefore, we do find a slight decrease in the line strengths of the $Fe^+$ lines for decreasing metal abundances, but the effect is very small. A rough glance at the spectrum may be very misleading.

## 12.4    Observed element abundances

### 12.4.1  *Population I stars*

With the curve of growth analysis as described in the previous chapter, we can determine abundances for all elements for which spectral lines can be identified and for which the necessary atomic data have been determined, either by laboratory experiments or by theoretical studies. We can study all stars which are bright enough to be observed with a dispersion high enough such that the different lines can be separated at least somewhere in the spectrum. When comparing stars of very different temperatures, the comparison of the abundances in the stars is usually accurate to within a factor of 2. When comparing stars of nearly equal temperatures and gravities, accuracies of about 25% can be reached in a very careful relative analysis. When higher accuracies are claimed this is usually misleading.

The stellar analyses show that almost all of the nearby stars have the same relative element abundances to within the uncertainty factor of about a factor of two. It is found that the vast majority of stars consists mainly of hydrogen with about 10% $\pm$ 5% of helium (by number of atoms and ions). Since each helium nucleus is four times as heavy as a proton, the abundance of helium by weight, called $Y$, can be computed from

$$Y = \frac{4 \times 0.1}{0.9 + 0.4} = 0.31\ (\pm 0.1).\qquad\qquad 12.27$$

Helium lines can only be seen in the spectra of hot stars. Therefore, the helium abundance can only be determined for these stars. In the cool stars

Table 12.3. *Abundances* log N *of the most abundant elements*[a]

| Atomic number | Element | Population I | | Population II | | |
|---|---|---|---|---|---|---|
| | | Sun[b] G2 V | α Lyr A0 V | γ Ser F6 IV | HD 140283 | HD 19445 |
| 1 | Hydrogen (H) | 12.0 | 12.0 | 12.0 | 12.0 | 12.0 |
| 2 | Helium (He) | (11.0) | (11.4) | ? | ? | ? |
| 8 | Oxygen (O) | 8.8 | 9.3 | 9.1 | ? | ? |
| 6 | Carbon (C) | 8.5 | ? | 8.4 | 6.4 | ? |
| 7 | Nitrogen (N) | 8.0 | 8.8 | ? | ? | ? |
| 10 | Neon (Ne) | (7.9) | ? | ? | ? | ? |
| 26 | Iron (Fe) | 7.6 | 7.1 | 7.2 | $5.2 \pm 0.3$ | $5.7 \pm 0.2$ |
| 14 | Silicon (Si) | 7.5 | 8.2 | 7.4 | 5.1 | 6.0 |
| 12 | Magnesium (Mg) | 7.4 | 7.7 | 7.5 | 4.9 | 6.5? |
| 16 | Sulfur (S) | 7.2 | ? | 7.2 | ? | ? |
| 13 | Aluminum (Al) | 6.4 | 5.7 | 6.1 | 3.7 | 4.5 |
| 28 | Nickel (Ni) | 6.3 | 7.0 | 6.4 | 4.2 | 4.7 |
| 20 | Calcium (Ca) | 6.3 | 6.3 | 5.9 | 4.0 | 4.8 |
| 11 | Sodium (Na) | 6.3 | 7.3 | 6.1 | 3.5 | ? |
| 24 | Chromium (Cr) | 5.9 | 5.6 | 4.9 | 3.6 | 3.8 |
| 17 | Chlorine (Cl) | 5.6 | ? | ? | ? | ? |
| 15 | Phosphorus (P) | 5.5 | ? | ? | ? | ? |
| 25 | Manganese (Mn) | 5.4 | 5.3 | 4.7 | 2.8 | 3.7 |
| 22 | Titanium (Ti) | 5.1 | 4.7 | 4.3 | 2.8 | 3.4 |
| 27 | Cobalt (Co) | 5.1 | ? | 4.1 | 2.7 | ? |
| 19 | Potassium (K) | 5.0 | ? | ? | ? | ? |

[a]Normalized to log $N(\mathrm{H}) = 12.00$.
[b]These abundances are considered to be 'cosmic abundances'.
Values in parentheses are uncertain.

we can see helium lines only in the chromospheric spectrum for which analysis is much more difficult and uncertain because LTE cannot be assumed. We have very incomplete knowledge about the helium abundance in cool stars and must resort to indirect methods which will be discussed in Volume 3. Unfortunately, the accurate knowledge of the helium abundance in old stars, which are all cool stars, is very important for many cosmological problems.

There is only a minor trace of heavy elements mixed in with the hydrogen and helium – only about $10^{-3}$ by number of atoms – but because these elements are heavy their fraction by mass, $Z$, still amounts to about 2%, i.e., $Z \approx 0.02$. This leaves for the hydrogen abundance by weight $X$

$$X = 0.67 \pm 0.1$$

for most of the 'normal' stars.

Among the heavy elements the most abundant ones are C, N, O, Ne. For each of these atoms or ions there are about 1000 hydrogen atoms or ions. The elements next in abundances are Fe, Si, Al, and Mg, but they are less abundant again by another factor of 10. Table 12.3 gives recent abundance determinations for the more abundant elements for stars in our neighborhood, the so-called Population I stars.

### 12.4.2   *Element abundances in Population II stars*

There are a few stars in our neighborhood, however, which show distinctly different abundances. They were originally noticed as subdwarfs (in the color magnitude diagram, they appear below the main sequence) or as weak lined stars. The stars generally also show large velocities relative to the sun and were therefore also called high velocity stars. We call these generally the Population II stars. These stars still mainly consist of hydrogen and helium, probably in almost the same proportions as in the Population I stars, but the heavy element abundances are generally lower by factors ranging from 3 to about 200. Surprisingly, the relative abundances within the group of heavy elements varies very little between Population I and Population II stars, although there is recent evidence that for intermediate degrees of depletion of the iron group elements the lighter elements are less depleted than the iron group elements. Also, variations in the ratio of the C, N, and O abundances are observed for various groups of stars. These details will be discussed in Volume 3, where we will study the origin of the chemical elements.

Element abundances for some high velocity Population II stars are also given in Table 12.3.

The depletion of heavy elements increases generally with increasing distance of the stars from the galactic plane. Those few high velocity stars in our neighborhood show by their velocities that they are actually on their way out to large distances in the halo again, from where they originally came. They just happen to pass by in their galactic orbit.

# 13

# Basics about non-local thermodynamic equilibrium

## 13.1 Einstein transition probabilities

In any equilibrium situation we expect that statistically the number of transitions per cm$^3$ each second into a given energy level should equal the number of transitions out of this energy level, so that the number of atoms in any given energy level does not change in time. This means we expect statistical equilibrium which must hold in any time independent situation. Let us look at any two energy levels in an atom which have the quantum numbers $l$ (lower level) and $u$ (upper level) (see Fig. 13.1). The number of atoms per cm$^3$ in the lower level is $N_l$ and in the upper level is $N_u$. The transition from level $l$ to level $u$ corresponds to an absorption process, where a photon of energy $h\nu_{l,u} = \chi_u - \chi_l$ is absorbed. The number of transitions per cm$^3$ is given by

$$n(l \to u) = N_l \bar{J}_{\nu(l,u)} B(l, u).\qquad 13.1$$

$B(l, u)$ is called the transition probability for the transition from $l$ to $u$; it is an atomic constant. (Some authors reverse the indices on the $B$, so one has to be careful.) $\bar{J}$ is the mean intensity averaged over the line. In the following we always mean $\bar{J}_{\nu(l,u)}$ if we write $J_{\nu(l,u)}$.

The number of spontaneous transitions from the upper level to the lower level is independent of the average light intensity and is given by

$$n(u \to l) = N_u A(u, l),\qquad 13.2$$

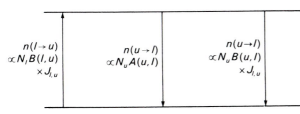

Fig. 13.1. For a given line we have to consider three kinds of radiative transitions, absorption (left), spontaneous emission (center), and induced emission (right).

where $A(u, l)$ is the transition probability for the spontaneous transitions from $u$ to $l$. It is also an atomic constant.

Generally, we do not know what the average intensity $J_{v(l,u)}$ is, but if we are dealing with thermodynamic equilibrium we know that

$$J_{v(u,l)} = B_{v(u,l)}(T),$$

where $B_v(T)$ is the Planck function for the temperature $T$. In the case of thermodynamic equilibrium we must require that for each transition there must be an equal number of radiative transitions going in both directions. We call this *detailed balancing*. In thermodynamic equilibrium, detailed balancing must be true because it can be shown that otherwise it would be possible to build a *perpetuum mobile* of the second kind, which we know is not possible. We also know that in thermodynamic equilibrium the ratio of the occupation numbers is governed by the Boltzmann formula, i.e.,

$$\frac{N_u}{N_l} = \frac{g_u}{g_l} e^{-(\chi_u - \chi_l)/kT}.$$  \hspace{2cm} 13.3

*If* these upward and downward transitions were the only processes that occur, the condition of detailed balancing would require that

$$n(u \rightarrow l) = n(l \rightarrow u),$$

which means

$$N_l \frac{2h\nu^3}{c^2} \frac{1}{e^{h\nu/kT} - 1} B(l, u) = N_u A(u, l)$$  \hspace{1cm} 13.4

or, using the Boltzmann formula,

$$\frac{2h\nu^3}{c^2} \frac{1}{e^{h\nu_{u,l}/kT} - 1} = \frac{g_u}{g_l} e^{-(\chi_u - \chi_l)/kT} \frac{A(u, l)}{B(l, u)}$$

$$= \frac{g_u}{g_l} e^{-h\nu_{u,l}/kT} \frac{A(u, l)}{B(l, u)} \qquad \text{since } h\nu_{u,l} = \chi_u - \chi_l. \quad 13.5$$

With $g_u$ and $g_l$ given and $A(u, l)$ and $B(l, u)$ atomic constants for each level, there is no way of determining these constants in such a way that this equation could be satisfied for all temperatures, as was first recognized by Einstein. An additional kind of transition is required which must depend also on the intensity of the light – the so-called induced emission processes. These are transitions from the upper to the lower level, whose number is proportional to the average intensity of the light $J_{v(u,l)}$. The number of those induced emissions per $cm^3$ s is described by

$$n'(u \rightarrow l) = N_u B(u, l) J_{v(u,l)}.$$  \hspace{2cm} 13.6

It can be shown that the photons emitted in this way due to the stimulation by a photon from the radiation field must have the same direction and phase as the stimulating photon. These induced emissions can therefore be treated as negative absorptions. With these additional processes the condition for detailed balancing in thermodynamic equilibrium now reads

$$N_l J_{v(u,l)}[B(l, u) - N_u J_{v(u,l)} B(u, l)] = N_u A(u, l). \qquad 13.7$$

Using $J_{v(u,l)} = B_{v(u,l)}$ and using the Boltzmann formula, we derive

$$\frac{2hv^3}{c^2} \frac{1}{e^{hv_{u,l}/kT} - 1} \left( B(l, u) \frac{g_l}{g_u} e^{hv_{u,l}/kT} - B(u, l) \right) = A(u, l). \qquad 13.8$$

This equation is satisfied if

$$B(u, l)g_u = B(l, u)g_l \qquad 13.9$$

and

$$A(u, l) = B(l, u) \frac{g_l}{g_u} \frac{2hv_{u,l}^3}{c^2} = B(u, l) \frac{2hv_{u,l}^3}{c^2}. \qquad 13.10$$

These are the well-known relations between the Einstein transition probabilities. Because the $B(u, l)$, $B(l, u)$, and $A(u, l)$ are atomic constants, these relations, although derived from thermodynamic equilibrium conditions, must always hold. We can therefore use them to get information for excitation conditions and the source function when we do not have thermodynamic equilibrium.

## 13.2 Induced emissions and lasers and masers

### 13.2.1 *Reduction of $\kappa_v$*

As we mentioned previously, the photons emitted in an induced emission process have the same directions and phases as the inducing photons. We can therefore treat the induced emission as a negative absorption.

The truly absorbed energy per $cm^3$ s is given by

$$\Delta E_v = - \iint dI_v \, d\omega \, dv = \iint \kappa_v I_v \, d\omega \, dv,$$

or

$$\Delta E_v = 4\pi \int \kappa_v J_v \, dv, \qquad 13.11$$

where the integral has to be extended over the whole line.

If we treat the induced emission as negative absorption, then the apparent net amount of energy absorbed can be described by

$$\Delta E_v' = \int_{\text{line}} \kappa_v' 4\pi J_v \, dv,$$

13.12

where $\kappa_v'$ is smaller than $\kappa_v$.

Remembering that each photon has an energy

$$E_{\text{ph}} = hv,$$

13.13

the net *number* of absorption processes per $cm^3$ s is given by

$$\frac{\Delta E_v'}{hv} = \int_{\text{line}} \frac{\kappa_v'}{hv} J_v 4\pi \, dv.$$

13.14

On the other hand, equation (13.1) tells us that the true number of absorbed photons is

$$n(l \to u) = N_l B(l, u) \bar{J}_v.$$

13.15

Comparing this with equation (13.14), we find

$$\int_{\text{line}} \frac{\kappa_v}{hv} J_v 4\pi \, dv = N_l B(l, u) \bar{J}_v,$$

13.16

and

$$\int_{\text{line}} \kappa_v \, dv = \frac{1}{4\pi} N_l B(l, u) hv,$$

13.17

where we have assumed that the $J_v$ within the line can be replaced by an average $\bar{J}_v$, and where the integrals have to be extended over the whole line.

For the apparent net number of absorption processes, we find similarly

$$n(l \to u) - n(u \to l) = N_l B(l, u) \bar{J}_v - N_u B(u, l) \bar{J}_v$$

$$= N_l B(l, u) \bar{J}_v \left( 1 - \frac{N_u B(u, l)}{N_l B(l, u)} \right) = \frac{4\pi}{hv} \bar{J}_v \int_{\text{line}} \kappa_v' \, dv. \quad 13.18$$

The comparison of equations (13.16) and (13.18) yields

$$\int_{\text{line}} \kappa_v' \, dv = \int_{\text{line}} \kappa_v \, dv \left( 1 - \frac{N_u B(u, l)}{N_l B(l, u)} \right).$$

13.19

If we now make use of equation (13.9), we derive

$$\kappa_v' = \kappa_v \left( 1 - \frac{N_u g_l}{N_l g_u} \right)$$

13.20

as the general relation between $\kappa_v'$ and $\kappa_v$.

If, as in thermodynamic equilibrium, the ratio $N_u/N_l$ is given by the Boltzmann formula for the kinetic temperature $T$, we find

$$\kappa_v' = \kappa_v (1 - e^{-hv/kT}),$$

13.21

since

$$\chi_u - \chi_l = h v_{u,l}.$$  13.22

For thermodynamic equilibrium conditions this is the relation between $\kappa'_\nu$ and $\kappa_\nu$. For other conditions we always have to go back to equation (13.20).

### 13.2.2 Lasers and masers

From equation (13.20) we see that, in principle, $\kappa'_\nu$ can become negative if

$$\frac{N_u}{N_l} \frac{g_l}{g_u} > 1.$$  13.23

As

$$dI_\nu = -\kappa'_\nu I_\nu \, ds,$$  13.24

we find for a negative $\kappa'_\nu$ that the intensity $I_\nu$ increases when the light beam passes through a gas with negative $\kappa'_\nu$. The original beam is intensified. This is what we find for a laser or maser. Under which circumstances can this happen? Clearly, we need

$$\frac{N_u}{N_l} > \frac{g_u}{g_l}.$$  13.25

For occupation numbers obeying the Boltzmann formula, this never happens. We need occupation numbers of the upper level which are much higher than those given by the Boltzmann formula. Such high occupation numbers of the upper level can be achieved by special pumping mechanisms, which we cannot discuss here.

## 13.3 The source function for a bound-bound transition

Our aim is to obtain information about the source function for a given transition between two energy levels with quantum numbers $u$ and $l$. The source function $S_\nu$ is defined as the energy $\varepsilon_\nu$ emitted per cm$^3$ s into the solid angle $d\omega = 1$, divided by the absorption coefficient per cm$^3$ $\kappa_\nu$, i.e.,

$$S_\nu = \varepsilon_\nu / \kappa_\nu.$$  13.26

In this expression, $\varepsilon_\nu$ includes the spontaneous transitions as well as the induced emission processes. If we include the induced emission processes as negative absorption and describe the net absorption by $\kappa'_\nu$, then we have to include in the emission $\varepsilon'_\nu$ only the spontaneous emission processes. We then have to express the source function as

$$S_\nu = \varepsilon'_\nu / \kappa'_\nu.$$  13.27

We now assume that the source function within each transition between two energy levels is independent of frequency, i.e., the source function in a given line is assumed to be frequency independent. This requires (see equation 13.27) that $\varepsilon'_v$ and $\kappa'_v$ have the same frequency dependence. We describe $\varepsilon'_v$ and $\kappa'_v$ by

$$\varepsilon'_v = \varepsilon'_0 \varphi(v) \quad \text{and} \quad \kappa'_v = \kappa'_0 \varphi(v), \qquad 13.28$$

with

$$\int_{\text{line}} \varphi(v)\, \mathrm{d}v = 1,$$

where the integral has to be extended over all frequencies within the line. We then find

$$S_v = \varepsilon'_0/\kappa'_0. \qquad 13.29$$

The net number of absorbed photons is

$$n(l \to u) - n(u \to l) = N_l \bar{J}_v B(l, u) - N_u B(u, l)\bar{J}_v = N_l \bar{J}_v B(l, u)\left(1 - \frac{N_u}{N_l}\frac{B(u, l)}{B(l, u)}\right)$$

$$= N_l \bar{J}_v B(l, u)\left(1 - \frac{N_u}{N_l}\frac{g_l}{g_u}\right) = 4\pi \frac{\bar{J}_v}{hv}\int_{\text{line}} \kappa'_v\, \mathrm{d}v = 4\pi \frac{\bar{J}_v}{hv}\kappa'_0. \qquad 13.30$$

Remembering again that the energy emitted per photon is $hv$, we find that the number of photons spontaneously emitted per cm$^3$ s is

$$\frac{4\pi}{hv}\int_{\text{line}} \varepsilon'_v\, \mathrm{d}v = \frac{4\pi}{hv}\varepsilon'_0 = N_u A(u, l). \qquad 13.31$$

Combining equations (13.30) and (13.31), we derive

$$S_v = \frac{\varepsilon'_0}{\kappa'_0} = \frac{N_u A(u, l)hv}{4\pi} \cdot \frac{4\pi}{N_l B(l, u)\left(1 - \frac{N_u g_l}{N_l g_u}\right)hv}. \qquad 13.32$$

Reordering the terms, we obtain

$$S_v = \frac{N_u}{N_l}\frac{A(u, l)}{B(l, u)}\frac{1}{\left(1 - \frac{N_u g_l}{N_l g_u}\right)}. \qquad 13.33$$

Using equation (13.10), we find

$$S_v = \frac{N_u}{N_l}\frac{g_l}{g_u}\frac{2hv^3}{c^2}\frac{1}{\left(1 - \frac{N_u g_l}{N_l g_u}\right)}. \qquad 13.34$$

After simplifying we obtain

$$S_v = \frac{2hv^3}{c^2} \frac{1}{\left(\dfrac{N_l g_u}{N_u g_l} - 1\right)}.$$  13.35

Note that in deriving this expression for the source function we have *not* made use of any equilibrium condition. This expression holds quite generally.

If the occupation numbers are given by the Boltzmann formula for the kinetic temperature $T$, then

$$S_v = \frac{2hv^3}{c^2} \frac{1}{e^{hv/kT} - 1} = B_v(T).$$  13.36

Here we have again made use of the relation $\chi_u - \chi_l = hv_{u,l}$.

We thus find that the source function equals the Planck function for the kinetic temperature; thus **LTE holds if the occupation numbers can all be described by the Boltzmann formula for the kinetic temperature**.

If this is not the case, i.e., if the occupation numbers cannot all be described by the Boltzmann formula for the kinetic temperature, then we can still describe the ratio of the occupation numbers for each pair of energy levels by a Boltzmann formula with some temperature $T_{exc}$, namely,

$$\frac{N_u}{N_l} = \frac{g_u}{g_l} e^{-(\chi_u - \chi_l)/kT_{exc}},$$  13.37

where this equation defines the excitation temperature $T_{exc}$. For a given ratio $N_u/N_l$, the $T_{exc}$ can always be calculated from equation (13.37). We then find that the source function is given by

$$S_v = \frac{2hv^3}{c^2} \frac{1}{(e^{hv/kT_{exc}} - 1)} = B_v(T_{exc}).$$  13.38

*For a given transition the source function is always given by the Planck function for the excitation temperature for the two levels involved in the transition.* In non-LTE (NLTE) conditions we will generally find different excitation temperatures for different pairs of energy levels.

The problem, of course, remains how to determine $N_u/N_l$ and $T_{exc}$ under NLTE conditions.

## 13.4 Excitation of energy levels

### 13.4.1 Collisional cross-sections

If we actually want to determine the ratio of the occupation numbers $N_u/N_l$, we have to look at the excitation mechanisms and at the equilibrium conditions. Let us start by looking at the excitation by collisions. Collisions with free electrons are usually most important.

The number of collisional excitations per $cm^3$ s is described by

$$n_c(l \to u) = N_l \int_0^\infty n_e(v) C'_{l,u}(v) \, dv = N_l n_e C_{l,u}(T).$$      13.39

Here, $C'_{l,u}(v)$ is the probability for an excitation of an electron in level $l$ into the level $u$ by a collision with a free electron with a velocity $v$. In order to get the total number of exciting collisions, we have to integrate over all collisions with electrons with different velocities. Under nearly all astrophysically important conditions the distribution of relative velocities corresponds to Maxwellian velocity distributions for a kinetic temperature $T$. The integral in equation (13.39) can therefore be described by $n_e C_{l,u}(T)$.

The number of collisional de-excitations from level $u$ to level $l$ is given by

$$n_c(u \to l) = N_u \int_0^\infty n_e(v) C'_{u,l}(v) \, dv = N_u n_e C_{u,l}(T).$$      13.40

In order to gain some insight into the relation between $C_{l,u}$ and $C_{u,l}$, we again make use of the well-known conditions in thermodynamic equilibrium. In thermodynamic equilibrium we must again have detailed balancing for the collisional processes. This means

$$n_c(u \to l) = n_c(l \to u),$$

or

$$N_l n_e C_{l,u}(T) = N_u n_e C_{u,l}(T),$$      13.41

which yields

$$\frac{N_u}{N_l} = \frac{C_{l,u}(T)}{C_{u,l}(T)},$$      13.42

where we have used equations (13.39) and (13.40). We know that in thermodynamic equilibrium the Boltzmann formula describes the ratio $N_u/N_l$, which then tells us that

$$\frac{g_u}{g_l} e^{-(\chi_u - \chi_l)/kT_{kin}} = \frac{C_{l,u}(T)}{C_{u,l}(T)}.$$      13.43

The ratio of the collisional cross-sections for excitation and de-excitation is then given by the left-hand side of this equation. Again, as long as the velocity distributions are Maxwellian, the collisional cross-sections are atomic properties depending only on the kinetic temperature. Their relations must be independent of the conditions of the radiation field. Equation (13.43) must therefore generally be true for Maxwell velocity distributions, which are usually established in astrophysical environments (see Bhatnagar, Krook, Menzel and Thomas, 1955).

## 13.4.2 Occupation numbers for collisional excitation and de-excitation

We can now proceed to determine the ratio $N_u/N_l$ for *statistical equilibrium* for the case of collisional excitation and de-excitation.

Under NLTE conditions we generally will not have detailed balancing for collisional processes, but will also have radiative excitations and de-excitations. Generally, we can only say that in equilibrium situations the sum of all excitation processes must equal the number of all de-excitation processes for a given energy level. For a real atom with many energy levels we have to consider the transitions into all other energy levels and from all other energy levels into the energy level under consideration. This leads to very long equations for each of the energy levels, and hence to laborious mathematics. The principles of the approach and of the results can well be demonstrated with the example of a hypothetical two-level atom. Therefore, we will restrict our discussion to this simple case, which will teach us the basic results without having to use laborious mathematics. However, the mathematics cannot be avoided when we want to determine the occupation numbers in real astrophysical cases.

For the present discussion we assume a very diluted radiation field such that the radiative processes are very rare and can be neglected in comparison with the collisional processes. We can also say that we assume a gas of fairly high density such that collisional processes are much more important than the radiative ones. The statistical equilibrium condition then requires

$$n_c(u \to l) = n_c(l \to u),$$

which leads to equation (13.41), however now we do not know $N_u/N_l$, but we can make use of equation (13.43). Inserting this into equation (13.41), we obtain

$$\frac{N_u}{N_l} = \frac{C_{l,u}(T)}{C_{u,l}(T)} = \frac{g_u}{g_l} e^{-(\chi_u - \chi_l)/kT_{kin}}. \qquad 13.44$$

**If the excitations and de-excitations are due to collisions, the occupation numbers follow the Boltzmann formula for the kinetic temperature.** We can conclude that in gases with high enough densities to make collisional excitations and de-excitations more important than the radiative processes, the occupation numbers follow the Boltzmann formula for the kinetic temperature. This means that the excitation temperature equals the kinetic temperature, which in turn means that the source function equals the Planck function for the kinetic temperature, **which means we have LTE.**

The question remains, what is the limit for a 'high density' gas? From our discussion it is clear that this limit must depend on the intensity of the radiation field and also on the radiative transition probabilities. If the radiative transition probabilities are very small (for instance, for the 21 cm line of hydrogen), then even the interstellar medium might qualify as a 'high density' gas. Generally, we may infer that the higher the intensity of the radiation field, the more important will be the radiative processes, and higher particle densities are required for the collisional processes to compete with the radiative excitations and de-excitations. In hot stars the radiation field is much more intense than in cool stars. Because of the higher absorption coefficients in hot stars, the gas pressures in the photospheres are lower than in cool stars. We can therefore expect that NLTE effects are more important in hot stars than in cool stars. We also expect that they are more important in low gravity – which means in high luminosity – stars than in main sequence stars. For these types of stars we have to study the special properties of the radiation field in order to see how important the NLTE effects actually are.

### 13.4.3  Occupation numbers for radiative excitation and de-excitation

If the excitation and de-excitation is determined by radiative processes only, the condition of statistical equilibrium requires again that the number of all excitations per $cm^3$ s must equal the number of all de-excitations per $cm^3$ s, which means

$$N_l B(l, u)\bar{J}_v - N_u B(u, l)\bar{J}_v = N_u A(u, l),$$  13.45

or

$$\bar{J}_v\left(\frac{N_l}{N_u} B(l, u) - B(u, l)\right) = A(u, l) = B(u, l)\frac{2hv^3}{c^2}.$$  13.46

Replacing $B(l, u)$ by $B(u, l)g_u/g_l$ and dividing by $B(u, l)$ gives

$$\bar{J}_v\left(\frac{N_l}{N_u}\frac{g_u}{g_l} - 1\right) = \frac{2hv^3}{c^2}.$$  13.47

From this expression it is obvious that for the case that

$$\bar{J}_v = B_v = \frac{2hv^3}{c^2}\frac{1}{e^{hv/kT} - 1}$$  13.48

we derive

$$\frac{N_l}{N_u} = \frac{g_l}{g_u}e^{-(\chi_u - \chi_l)/kT},$$  13.49

which is the Boltzmann formula. Even if excitation and de-excitation are due to radiative processes only, deviation from LTE occurs only if $\bar{J}_v$ deviates from $B_v$.

In the general case we derive for radiative excitation and de-excitation only that

$$\frac{N_l}{N_u}\frac{g_u}{g_l} - 1 = \frac{2h\nu^3}{c^2}\frac{1}{J_\nu},$$

13.50

or

$$\left(\frac{N_l g_u}{N_u g_l} - 1\right)\Big/(e^{h\nu/kT_{kin}} - 1) = \frac{2h\nu^3}{c^2}\frac{1}{(e^{h\nu/kT_{kin}} - 1)}\frac{1}{\bar{J}_\nu} = \frac{B_\nu}{\bar{J}_\nu}.$$

13.51

The left-hand side of this equation gives essentially the ratio of the actual occupation numbers to the one obtained from the Boltzmann formula, which is determined by the ratio of $B_\nu/\bar{J}_\nu$. However, this is true only for the two level atom, discussed here.

## 13.5   Summary

In summary, we may conclude that large deviation from LTE may be expected for low density gas in which the radiation field deviates strongly from the Planck function for the kinetic temperature. Such deviations occur in hot stars due to the discontinuities in the continuous absorption coefficients, especially in the ultraviolet, where most of the flux is emitted. Large deviations are also expected in the low density chromospheres and coronae of cool stars, where the radiation field corresponds to a diluted Planck function for the effective temperature of the star, while the kinetic temperature in the coronae may be several million degrees.

In the photospheres of cool stars, in which the densities are rather high and for which the radiation field does not deviate much from the Planck function, we may expect that the LTE approximation will be a good one, and that our discussions of the photospheric structure will not suffer from this approximation.

Actual calculations – mainly by Mihalas (e.g., 1978) and collaborators – have shown that the LTE approximation is not too bad for all main sequence stars and giants with temperatures less than about 25 000 K. For main sequence O stars it does not seem to be a valid approximation. We also have to check this assumption for supergiants.

# 14

---

# The hydrogen convection zone

## 14.1 Introduction

In Volume 1 we saw that the surface of the sun is not smooth, but that we see bright granules separated by darker intergranular lanes. The structures appear to have dimensions of the order of 500 km diameter and therefore can only be recognized under conditions of very good seeing (1 arcsec corresponds to 700 km on the sun). There may be even smaller structures which we cannot resolve because of the atmospheric seeing and because the solar observation satellites so far do not have mirrors large enough to resolve such small-scale structures. When we discussed these structures on the solar surface, we pointed out that in the bright regions the motions, measured by the Doppler shift, are mainly directed outwards, while in the dark intergranulum the motions are mainly downwards. These motions and temperature inhomogeneities seen in the granulation pattern are due to the hydrogen convection zone just below the solar photosphere. These motions in the hydrogen convection zone are believed to be the source of the mechanical energy flux which heats the solar chromosphere and corona. Similar convection zones in other stars are believed to be responsible for the heating of the stellar chromospheres and coronae whose spectra are observed in the ultraviolet and in the X-ray region by means of satellites. Before we can discuss these outer layers of the stars, we have to discuss briefly the reason for these convection zones and the velocities expected to be generated by this convection, and how these and the mechanical energy flux generated by these motions are expected to vary for different stars with different $T_{eff}$, gravity, and with different chemical abundances.

## 14.2 The Schwarzschild instability criterion

We are all familiar with the phenomenon of convection, since our daily weather is caused by convection in the Earth's atmosphere. On a

174

smaller scale, we can witness convection on a hot summer day when the asphalt of a highway is heated by the solar radiation and the hot air rises above the asphalt, causing the light beams to be bent due to refraction in the different temperature regions, showing us the flimmering image of the road. We see the same phenomenon above a hot radiator. We always see these convective motions if we have a large temperature gradient in a fluid or gas. How large does the temperature gradient have to be in order to cause convection? This was first studied quantitatively by Karl Schwarzschild. We discuss here his results, which can best be understood when looking at Fig. 14.1. Here we have plotted schematically the temperature stratification in a stellar atmosphere as a function of the geometrical depth $t$. Of course, we could also use the gas pressure $P_g$ as the abscissa. Suppose that at the point $P_1$ a small gas bubble rises accidentally by a small amount. This small shift upwards brings it into a layer with a slightly lower gas pressure. The bubble therefore expands and in the process cools off adiabatically if it does not have any energy exchange with the surroundings. Therefore, it experiences a temperature change given by

$$\left(\frac{dT}{dt}\right)_{bubble} = \frac{dP_g}{dt}\left(\frac{dT}{dP_g}\right)_{adiabatic}$$

14.1

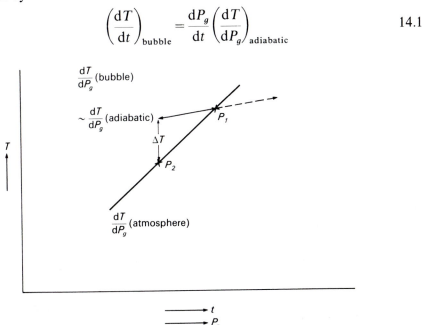

Fig. 14.1. A gas bubble accidentally rises from $P_1$ to the point $P_2$. At the higher layer it expands to the lower gas pressure of the new surroundings and cools off nearly adiabatically. At $P_2$ it still has a higher temperature than the surroundings if the adiabatic temperature gradient $dT/dP_g$ is smaller than the corresponding temperature gradient in the surrounding atmosphere.

The question is now, will this gas bubble continue to rise? Or will it fall back to its original position? If it continues to rise, then the atmosphere will soon move everywhere, because further small disturbances are created by the rising bubble; the atmosphere is then unstable against convection. On the other hand, if the bubble experiences a force that pushes it back to its starting point, then no motion can be kept up. The atmosphere is stable against convection.

The only force accelerating a floating gas bubble is the buoyancy force. If the bubble has a lower density than its surroundings, then the buoyancy force will push it upwards and it will continue to rise. If the density of the bubble is higher than in the surroundings, the bubble will fall back. Because the bubble expands to pressure equilibrium with the surroundings it has a lower density than the surrounding gas only if it has a higher temperature than the surrounding gas. According to equation (14.1), the temperature of the gas bubble at point $P_2$ is given by

$$T(\text{bubble}) = T_1 + \left(\frac{dT}{dt}\right)_{\text{bu}} \Delta t = T_1 + \frac{dP_g}{dt}\left(\frac{dT}{dP_g}\right)_{\text{ad}} \Delta t. \qquad 14.2$$

$T_1$ is the temperature at point $P_1$. $\Delta t$ is $t(P_2) - t(P_1)$ (bu means bubble, ad means adiabatic). The temperature of the surrounding gas is determined by the average temperature stratification in the atmosphere, which, in the absence of convection, is determined by radiative equilibrium, as discussed in Chapter 5. At the point $P_2$ the temperature of the surrounding gas is therefore given by

$$T(P_2) = T_1 + \frac{dT}{dt_{\text{bu}}} \Delta t = T_1 + \frac{dP_g}{dt}\frac{dT}{dP_{g\,\text{at}}} \Delta t \qquad 14.3$$

(at means surrounding atmosphere). Subtraction of equation (14.3) from equation (14.2) shows that

$$T(\text{bubble}) - T(P_2) = \frac{dP_g}{dt}\left(\frac{dT}{dP_{g\,\text{ad}}} - \frac{dT}{dP_{g\,\text{at}}}\right)\Delta t. \qquad 14.4$$

(Remember, for the rising bubble $\Delta t < 0$.) The temperature of the rising bubble is larger than the temperature of the surrounding if the adiabatic gradient $dT/dP_{g\,\text{ad}}$ is less than the temperature gradient in the surrounding gas; i.e., the **condition for convective instability is**

$$\frac{dT}{dP_{g\,\text{ad}}} < \frac{dT}{dP_{g\,\text{at}}}. \qquad 14.5$$

If we multiply both sides of the inequality (14.5) with the positive value $P_g/T$, we can write the same inequality as

$$\left(\frac{d \log T}{d \log P_g}\right)_{ad} < \left(\frac{d \log T}{d \log P_g}\right)_{at}. \qquad 14.6$$

It has become customary to abbreviate the logarithmic gradient by

$$\frac{d \log T}{d \log P_g} = \nabla. \qquad 14.7$$

If we start out with an atmosphere in radiative equilibrium with $\nabla = \nabla_{rad}$, then convection sets in if

$$\nabla_{ad} < \nabla_{rad}. \qquad 14.8$$

We shall soon see that equation (14.6) always holds if equation (14.8) holds. Equation (14.8) therefore is generally the condition for convective instability in a stellar atmosphere, provided that the mean atomic weight $\mu$ is constant in the layer under consideration. Without saying so, we have used the assumption of constant $\mu$ when stating that the density is lower only when the temperature of the bubble is higher than in the surroundings. A changing $\mu$, due to ionization, can easily be taken into account. A change in $\mu$ due to different chemical abundances may be very difficult to take into account, as we shall see in Volume 3.

For our purposes here, condition (14.8) is sufficiently accurate to determine convective instability.

From Fig. 14.1, it is quite obvious that if condition (14.8) is satisfied and the upwards displaced bubble keeps rising, then a downward displaced bubble has a higher density than the surrounding gas because the falling bubble has a lower temperature than the surroundings. Falling motions therefore also continue.

## 14.3    Reasons for convective instability

### 14.3.1    *The radiative temperature gradient*

When looking at equation (14.8) it is obvious that there may be in principle two reasons for convective instability: either $\nabla_{ad}$ becomes very small, or $\nabla_{rad}$ becomes very large. We shall first follow the second possibility. In order to check when the radiative temperature gradient will become very large, we have to calculate it first in order to see how it depends on various parameters.

For radiative equilibrium we formerly found

$$\frac{4}{3\bar{\kappa}_{cm}}\frac{dB}{dt} = \frac{4}{3}\frac{dB}{d\tau} = F_r = \frac{4}{3}\frac{4\sigma}{\bar{\kappa}_{cm}\pi}T^3\frac{dT}{dt} \qquad \text{because } B = \frac{\sigma}{\pi}T^4. \qquad 14.9$$

Solving for d ln $T/dt$, we obtain

$$\frac{d \ln T}{dt} = \frac{\pi F_r 3 \bar{\kappa}_{cm}}{16 \sigma T^4}.$$  14.10

In order to calculate d ln $P/dt$, we have to make use of the hydrostatic equation

$$\frac{d \ln P_g}{dt} = g \frac{\rho}{P_g}.$$  14.11

Division of equation (14.9) by equation (14.10) yields

$$\nabla_{rad} = \frac{d \ln T}{dt} \bigg/ \frac{d \ln P_g}{dt} = \frac{3 \pi F_r \bar{\kappa}_{gr} P_g}{16 \sigma T^4 g}.$$  14.12

From this equation we see that there are two ways in which $\nabla_{rad}$ may become very large:

1. If $F_r$ becomes very large. This may happen close to the center of the star if all the luminosity of the star is generated very close to the center and has to get through a very small surface area $4 \pi r^2$. This leads to a central convection zone in hot stars.
2. If $\bar{\kappa}_{gr}$ becomes very large, while $P_g$ also remains large. This only happens if $\bar{\kappa}_{gr}$ increases steeply with depth, as we shall see now.

The hydrostatic equation says

$$\frac{dP_g}{d\bar{\tau}} = \frac{g}{\bar{\kappa}_{gr}}.$$  14.13

If $\bar{\kappa}_{gr}$ is constant with depth, then integration of this equation gives

$$P_g = \frac{1}{\bar{\kappa}_{gr}} \bar{\tau} g,$$  14.14

which means

$$\bar{\kappa}_{gr} P_g = g \bar{\tau}.$$

With $T^4 \propto \bar{\tau}$ we see from equation (14.12) that $\nabla_{rad}$ becomes independent of $T^4$ and $P_g$ as well as of $\bar{\tau}$, because $T^4$ is proportional to $\bar{\tau}$ for $\bar{\tau} \gg \frac{2}{3}$.

If we now approximate the depth dependence of $\bar{\kappa}_{gr}$ (see equation 9.14) by

$$\bar{\kappa}_{gr} = A P_g^b$$  14.15

then integration of the hydrostatic equation (14.13) yields

$$\frac{1}{b+1} P_g^{b+1} = \frac{g}{A} \bar{\tau}; \quad \text{and} \quad \frac{d \ln P_g}{d\bar{\tau}} = \frac{g}{A} \frac{1}{P_g^{b+1}} = \frac{1}{(b+1)\bar{\tau}}.$$  14.16

The radiative temperature gradient can now be written as (see equation 14.10)

$$\frac{d \ln T}{d\bar{\tau}} = \frac{\pi F_r 3}{16 \sigma T^4} = \frac{1}{4} \frac{1}{(\bar{\tau} + \frac{2}{3})}.$$  14.17

Here we have used the radiative equilibrium temperature stratification for a grey atmosphere, namely

$$\sigma T^4 = \tfrac{3}{4}\sigma T_{\text{eff}}^4(\bar{\tau} + \tfrac{2}{3}) = \tfrac{3}{4}\pi F_r(\bar{\tau} + \tfrac{2}{3}). \qquad 14.18$$

Division of equation (14.17) by equation (14.16) yields

$$\nabla_{\text{rad}} = \frac{d \ln T}{d\bar{\tau}} \bigg/ \frac{d \ln P_g}{d\bar{\tau}} = \frac{(b+1)\bar{\tau}}{4(\bar{\tau} + \tfrac{2}{3})}. \qquad 14.19$$

We now see that $\nabla_{\text{rad}}$ does not depend on $A$, which means it does not depend on the absolute value of $\bar{\kappa}$, but only on the increase of $\bar{\kappa}$ with depth described here by the exponent $b$ for the dependence of $\bar{\kappa}$ on $P_g$. $\nabla_{\text{rad}}$ becomes large only for large values of $b$. At the surface the $\bar{\kappa}$ needs to be small such that a large gas pressure is obtained. If $\bar{\kappa}$ then increases towards deeper layers the product $\bar{\kappa} P_g$ can become large because $P_g$ remains large even when $\bar{\kappa}$ increases.

A look at Fig. 8.7 shows that $\bar{\kappa}$ increases steeply for temperatures above 6000 K. At these temperatures the hydrogen just barely starts to ionize. Because hydrogen is so abundant, even 0.1% ionization of hydrogen increases the number of free electrons by a factor of 10. The increase in the electron density leads to an increase in the $H^-$ absorption coefficient. At the same time the hydrogen absorption coefficient also increases because of the increasing excitation of the third and second energy levels of the hydrogen atom. For somewhat higher temperatures than 6000 K this latter effect becomes more important than the increase in the $H^-$ absorption.

### 14.3.2 *The adiabatic temperature gradient*

Let us now see under which circumstances the adiabatic gradient can become very small.

By definition, an adiabatic change in pressure and temperature means a change with no heat exchange with the surroundings, i.e.,

$$dQ = 0 \qquad \text{with } dQ = dE + P_g\, dV, \qquad 14.20$$

where $E$ is the internal energy. This is

$$dE = C_v\, dT, \qquad 14.21$$

where $C_v$ is the specific heat per mol at constant volume, if we consider 1 mol of gas. For an ideal gas we have the equation of state

$$P_g V = R_g T, \qquad 14.22$$

where $V$ is the volume of 1 mol of gas, i.e., $V = \mu/\rho$. Differentiation of this equation yields

$$P_g\, dV + V\, dP_g = R_g\, dT, \qquad 14.23$$

or

$$dT = \frac{P_g \, dV + V \, dP_g}{R_g}. \tag{14.24}$$

For constant pressure we obtain with $dP_g = 0$

$$P_g \, dV = R_g \, dT. \tag{14.25}$$

The specific heat $C_p$ per mol is defined as the amount of energy that has to be transferred to the gas at constant pressure to increase the temperature by one kelvin, i.e., for $\Delta T = 1$ K.

For $\Delta T = 1$ we find, using equations (14.20) and (14.25),

$$\Delta Q = C_p = C_v + R_g, \tag{14.26}$$

or

$$C_p - C_v = R_g.$$

Let us now consider adiabatic changes with $dQ = 0$. This means

$$dE + P_g \, dV = 0 \qquad \text{where } dE = C_v \, dT. \tag{14.27}$$

Using equations (14.24) and (14.26), we derive

$$dT = \frac{P_g \, dV + V \, dP_g}{C_p - C_v}. \tag{14.28}$$

Inserting this into equation (14.27) we obtain

$$\frac{C_v}{C_p - C_v} (P_g \, dV + V \, dP_g) + P_g \, dV = 0. \tag{14.29}$$

After dividing by $P_g V$ and multiplying by $(C_p - C_v)/C_v$, we find

$$\frac{dV}{V} \left( 1 + \frac{C_p - C_v}{C_v} \right) + \frac{dP_g}{P_g} = 0, \tag{14.30}$$

or

$$\frac{C_p}{C_v} \, d \ln V = -d \ln P_g. \tag{14.31}$$

Introducing $\gamma = C_p/C_v$ and integrating equation (14.31) gives

$$\gamma \ln \frac{V}{V_0} = -\gamma \ln \frac{\rho}{\rho_0} = -\ln \frac{P_g}{P_{g0}}. \tag{14.32}$$

After taking the exponential we derive at

$$\left( \frac{\rho}{\rho_0} \right)^{\gamma} = \frac{P_g}{P_{g0}} \quad \text{or} \quad P_g = P_{g0} \left( \frac{\rho}{\rho_0} \right)^{\gamma} \quad \text{or} \quad P_g \propto \rho^{\gamma}. \tag{14.33}$$

With this relation we can now easily determine $\nabla_{\text{ad}}$. For an ideal gas $P_g = R_g T \rho / \mu$ and $d \ln P_g = d \ln T + d \ln \rho$ or

$$1 = \frac{d \ln T}{d \ln P_g} + \frac{d \ln \rho}{d \ln P_g}. \tag{14.34}$$

For an adiabatic change we know from equation (14.32) that

$$\frac{d \ln \rho}{d \ln P_g} = \frac{1}{\gamma} = \frac{C_v}{C_p}. \qquad 14.35$$

If we insert this into equation (14.34), we obtain

$$\left(\frac{d \ln T}{d \ln P_g}\right)_{ad} = \nabla_{ad} = 1 - \frac{1}{\gamma} = \frac{\gamma - 1}{\gamma}. \qquad 14.36$$

For a monatomic gas the internal energy per particle is generally $\frac{3}{2}kT$, and for a mol it is $\frac{3}{2}R_g T$, and therefore

$$\frac{dE}{dT} = C_v = \frac{3}{2}R_g. \qquad 14.37$$

Since, according to equation (14.26) $C_p = C_v + R_g$ we have

$$C_p = \frac{5}{2}R_g \quad \text{and} \quad \gamma = \frac{C_p}{C_v} = \frac{5}{3}. \qquad 14.38$$

With equation (14.36) we therefore find

$$\nabla_{ad} = \frac{2}{3}/\frac{5}{3} = \frac{2}{5} = 0.4 \qquad 14.39$$

for a monatomic gas with $E = \frac{3}{2}R_g T$.

Equation (14.36) tells us that $\nabla_{ad}$ will become very small if $\gamma$ approaches 1. Since

$$\gamma = \frac{C_p}{C_v} = \frac{C_v + R_g}{C_v} \quad \text{or} \quad \gamma = 1 + \frac{R_g}{C_v} \qquad 14.40$$

$\gamma$ approaches 1 if $C_v$ becomes very large. Remembering that $C_v$ is the amount of heat energy needed to heat one mol of gas by one degree kelvin, this means if a large amount of energy is needed to heat the gas by one degree, $\gamma$ will approach 1 and the adiabatic gradient becomes very small. In other words, if such a gas expands there is a large amount of energy available for the expansion before the gas cools by one degree. Under those circumstances, $\nabla_{ad}$ becomes very small and convection sets in. We find such a situation for temperatures and pressures for which an abundant element like hydrogen or helium starts to ionize. In this case, a large amount of energy is used to remove additional electrons from the hydrogen or helium atom or ion as needed for a temperature increase according to the Saha equation. This energy is not available to increase the temperature, i.e., the kinetic energy of the average particle. Large amounts of energy are necessary to heat the gas and at the same time supply the necessary energy for the increased ionization.

Depending on the gas pressure, hydrogen starts to ionize for temperatures between 6000 and 7000 K in the stellar atmospheres. For higher temperatures $\nabla_{ad}$ is therefore smaller than 0.4 and may be as small as 0.05.

In Section 14.3.1 we found that for $\tau \geqslant 1$ roughly $\nabla_{rad} = \nabla_r = (b+1)/4$. For constant $\bar{\kappa}$, i.e., $b=0$, convection sets in for $\nabla_{ad} \leqslant 0.25$, which is easily obtained in the hydrogen or helium ionization zones. This decrease in $\nabla_{ad}$ occurs in the same region where we find the increase in the absorption coefficient which is coupled with the beginning ionization of hydrogen. Large $\nabla_r$ and small $\nabla_{ad}$ therefore occur in the same temperature range and lead to one convection zone, the so-called *hydrogen convection* zone, because it is due to the ionization of hydrogen. $\nabla_{ad}$ also becomes small when helium ionizes at somewhat higher temperatures than hydrogen. We call this unstable region the helium convection zone. For most stars these zones merge. On the main sequence we find only for early F stars separate hydrogen and helium convection zones.

## 14.4    The upper boundaries for the hydrogen convection zones

As we saw earlier (see Chapters 8 and 9), we calculate the temperature and pressure stratification in a stellar atmosphere by integration of the radiative equilibrium equation, which determines the temperature stratification, and by stepwise integration of the hydrostatic equation, which determines the pressure stratification. At each step we can check whether $\nabla_{ad}$ is smaller than $\nabla_r$ and determine where convective instability sets in. In Fig. 14.2 we show the optical depths for the upper boundaries of the hydrogen convection zones as a function of the $T_{eff}$ for stars with different gravities, i.e., with different luminosities. Because

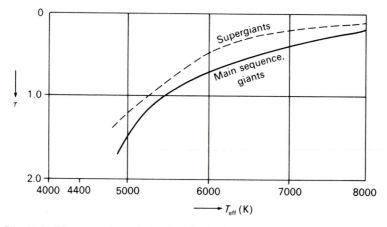

Fig. 14.2. The upper boundaries for the hydrogen convection zones are shown as a function of the effective temperatures of the stars. The different curves refer to stars with different gravitational accelerations $g$ (and thus different luminosities).

convection sets in essentially at a given temperature, the convection sets in at lower optical depths for stars with higher effective temperatures. When the effective temperature is so high that hydrogen is already ionized at the surface of the star, there is no longer a steep increase in $\bar{\kappa}$ and hydrogen convection stops. There are still the helium convection zones, but they are of little importance for the stellar atmospheres.

## 14.5 Convective energy transport

Once convection sets in, hot gas is transported to the surface and cool gas is traveling to the bottom. This means convection transports energy. This energy transport modifies the energy equilibrium, which is very important for the calculation of the temperature stratification in the interiors of the star. At this point we are more interested in the effects of convection at the surface. Nevertheless, an estimate of the convective energy transport in the upper parts of the convection zone will enable us to give a rough estimate for the production of mechanical flux available to heat the chromospheres and coronae. Let us try to calculate how much heat energy is transported by convection through a horizontal layer at depth $t_0$ in Fig. 14.3. At this layer hot bubbles come from below, having started at different depths. Cool bubbles come from above, also having started at different layers. The temperature differences between the bubbles and the surroundings depends on the starting point of the bubbles. In the following we will generally consider average temperature differences $\overline{\Delta T}$ at any given layer and also average velocities $\bar{v}$ at any given layer. The net heat energy transport through 2 cm$^2$ of a horizontal layer at $t_0$ is given by the heat transported upwards minus the heat transported down by the downflow of material (see Fig. 14.4). Through 1 cm$^2$ the material is flowing upwards; through 1 cm$^2$ the material is flowing downward. The total mass flow must, of course, be zero; otherwise, the material would either have to collect at the bottom or at the top, which would upset the pressure equilibrium. The total mass flow upward per second is

$$m_u = \rho_u \bar{v}_u, \qquad \rho = \text{density}, \quad v = \text{velocity} \qquad 14.41$$

where the index $u$ stands for upwards.

Similarly, the mass flow downward is given by

$$m_d = \rho_d \bar{v}_d, \qquad 14.42$$

where the index $d$ stands for downwards. The heat content per cm$^3$ is given by

$$Q = \rho c_p T. \qquad 14.43$$

Here $\dot{c}_p$ is the specific heat per gram at constant pressure. Since we compare the heat contents of the upward and downward moving material, which we consider to be at pressure equilibrium, we have to use the specific heat at constant pressure. Each second a column of length $v$ is flowing through a horizontal layer. The net heat transport through 2 cm$^2$ at $t_0$ is then

$$2\pi F_c = \rho_u \bar{v}_u c_p \overline{T_u} + \tfrac{1}{2}\rho_u \bar{v}_u^3 - \rho_d \bar{v}_d c_p \overline{T_d} - \tfrac{1}{2}\rho_d \bar{v}_d^3. \qquad 14.44$$

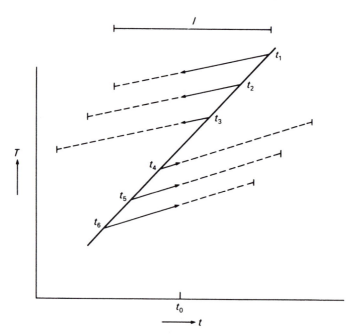

*Fig. 14.3.* At a given horizontal layer at depth $t_0$ bubbles coming from above fall through this layer and bubbles coming from below rise through this layer. They have started at different depths $t_1, t_2, t_3, t_4, t_5$ and therefore at $t_0$ have different temperature differences with respect to the average surroundings. For the convective energy transport we consider the average temperature difference $\overline{\Delta T}$ corresponding to an average path length $l/2$ for bubbles passing through $t_0$, when $l$ is the average total path length for the rising or falling gas.

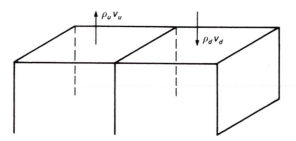

*Fig. 14.4.* At a given horizontal layer gas is rising through half of the area and is falling through the other half of the area. When calculating the convective energy transport through 2 cm$^2$ of the horizontal layer we consider upward moving gas in 1 cm$^2$ and downward moving gas in the other 1 cm$^2$.

Here we have added the transport of kinetic energy. $\pi F_c$ is the convective energy transport per cm$^2$.

We now have to consider the mass transport, which has to be zero, as we stated above. This means

$$\rho_u \bar{v}_u = \rho_d \bar{v}_d \quad \text{which for} \quad \rho_u \approx \rho_d \approx \rho \quad \text{yields} \quad \bar{v}_u = \bar{v}_d = \bar{v}. \quad 14.45$$

At this point we can assume that the densities are nearly the same, since it is not the difference of the densities which is used here. Inserting this into equation (14.44) and reordering, we obtain

$$\rho \bar{v} c_p (\overline{T_u} - \overline{T_d}) + \tfrac{1}{2} \rho \bar{v} (\bar{v}_u^2 - \bar{v}_d^2) = 2\pi F_c. \quad 14.46$$

With $v_u \approx v_d$ the second term on the left-hand side becomes zero, which means the kinetic energy transport does not contribute to the energy transport because the contributions from the upwards and downwards flowing material nearly cancel. This is due to the requirement of zero net mass transport. We are then left with

$$\pi F_c = \tfrac{1}{2} \rho \bar{v} c_p (\overline{T_u} - \overline{T_d}). \quad 14.47$$

With $\overline{\Delta T}$ being the average temperature difference between the upwards flowing material and the average surrounding, we have

$$\overline{T_u} - \overline{T_d} = 2\overline{\Delta T}, \quad 14.48$$

and

$$\pi F_c = \rho \bar{v} c_p \overline{\Delta T}. \quad 14.49$$

The fact that in equation (14.47) the difference $\overline{T_u} - \overline{T_d}$ enters and not only $\overline{\Delta T}$ tells us that the downward moving material carrying negative energy downward contributes as much to the convective energy transport as the upwards moving material carrying positive energy upward.

Including the convective energy transport in our thermal equilibrium condition $\pi F = \text{const.} = \sigma T_{\text{eff}}^4$ (see equation 6.9), we must now require that

$$\pi F = \pi F_r + \pi F_c = \sigma T_{\text{eff}}^4. \quad 14.50$$

With $\pi F_r \geqslant 0$ we find that

$$\pi F_c = \rho \bar{v} c_p \overline{\Delta T} \leqslant \sigma T_{\text{eff}}^4. \quad 14.51$$

This equation gives us an upper limit for the velocities $\bar{v}$ once we know the relation between $\overline{\Delta T}$ and $\bar{v}$. That such a relation must exist becomes obvious when we remember that the $\overline{\Delta T}$ determine the buoyancy force which gives the acceleration. If $f_b$ is the buoyancy force per cm$^3$, we have

$$f_b = \Delta \rho g, \quad 14.52$$

where $g$ is the gravitational acceleration. The acceleration $a$ due to the buoyancy force is then

$$a = \frac{\Delta \rho}{\rho} g = \frac{\text{force}}{\rho}. \quad 14.53$$

Assuming again pressure equilibrium between the bubble and the surrounding gas, we find (assuming constant $\mu$)

$$\frac{\Delta\rho}{\rho} = -\frac{\Delta T}{T} + \frac{\Delta P}{P} = -\frac{\Delta T}{T},$$    14.54

and

$$f_b = -\frac{\Delta T}{T}\rho g.$$    14.55

With energy $= \int$ force d$t$ we find

$$\tfrac{1}{2}\rho v^2 = \int_{t_0}^{t}\frac{\Delta T}{T}\rho g\,\mathrm{d}t \quad\text{and}\quad v^2 = 2\int_{t_0}^{t}\frac{\Delta T}{T}g\,\mathrm{d}t.$$    14.56

From Fig. 14.1 and our discussion in Section 14.3.2 we see that

$$\Delta T = \left[\left(\frac{\mathrm{d}T}{\mathrm{d}t}\right)_{\mathrm{atm}} - \left(\frac{\mathrm{d}T}{\mathrm{d}t}\right)_{\mathrm{ad}}\right]\Delta t = \Delta\,\mathrm{grad}\,T\cdot\Delta t.$$    14.57

Introducing this into equation (14.56), we derive

$$v^2(\Delta t) = 2g\int_{0}^{\Delta t}\frac{\Delta\,\mathrm{grad}\,T}{T}\Delta s\,\mathrm{d}s = 2g\,\frac{\Delta\,\mathrm{grad}\,T}{T}\tfrac{1}{2}\Delta t^2.$$    14.58

Here we have assumed that $\Delta\,\mathrm{grad}\,T$ remains constant over the path length of the bubble. If this is not so we have to use an average value, averaged over this path length.

For the energy transport through a given layer, we have to use an average $v$ at this layer. In order to find an average velocity we have to use an average path length in equation (14.58). We call this average path length for bubbles passing through $t_0$ the $\overline{\Delta t} = \tfrac{1}{2}l$. $l$ is usually referred to as the mixing length. We want to emphasize that here $l$ refers only to the average path length for the rising or falling gas.

Introducing this average path length $l$ into equation (14.58) for the average $\Delta T$ we find

$$\bar{v}^2 = g\,\frac{\Delta\,\mathrm{grad}\,T}{T}\left(\frac{l}{2}\right)^2 \quad\text{and}\quad \overline{\Delta T} = \Delta\,\mathrm{grad}\,T\frac{l}{2},$$    14.59

showing that

$$\overline{\Delta T} = \bar{v}^2\,\frac{T}{g}\frac{2}{l}.$$    14.60

With this we find from our equation (14.51) that

$$\pi F_c = \rho c_p\,\frac{T}{g}\frac{2}{l}\,v^3 \leqslant \sigma T_{\mathrm{eff}}^4,$$    14.61

which now gives us an upper limit for $v^3$. If at any point in the convection zone essentially all the energy is transported by convection, then this upper limit for $\bar{v}^3$ does indeed give the true velocity at that layer in the atmosphere. Equation (14.61) thus provides a simple way of calculating the maximum convective velocity in a given convection zone.

In Figs. 14.5 and 14.6 we show temperature differences and velocities calculated for stars with different effective temperatures and gravities.

## 14.6 Importance of convective energy transport

From equations (14.51) and (14.61) we see that the convective energy transport is proportional to the density $\rho$. Higher in the atmosphere we therefore expect the convective energy transport to decrease. In the photospheric parts of the convection zones in F stars we also expect very small contributions of convective energy transport to the total energy transport. In Fig. 14.7 we show the optical depths for which convection contributes 1% to the total energy transport. In the photospheres, which means for $\tau \leqslant 1$, convective energy transport is never very important. We should, however, add a note of caution. As we see from equation (14.61), rather high velocities and temperature differences are expected in stars with low density convection zones, for instance in main sequence F stars or in giant G stars, if in the deeper layers convective energy transport is important. These temperature inhomogeneities do prevail to the surface layers and can cause rather noticeable effects on the spectra and the colors of these stars.

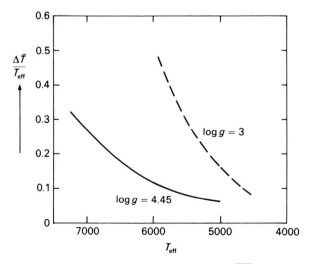

*Fig. 14.5.* Maximum temperature differences $\overline{\Delta T}$ between rising or falling gas and the average temperature of the surrounding are shown as a function of the effective temperatures of the stars for main sequence stars.

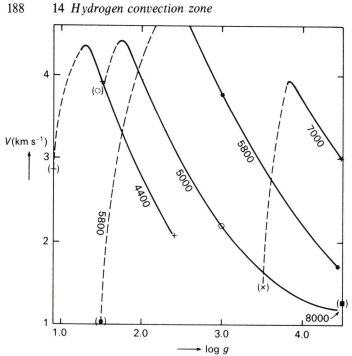

Fig. 14.6. Average convective velocities of the rising or falling gas are shown as a function of the effective temperatures of the stars for stars with different gravities.

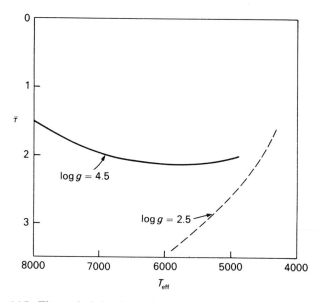

Fig. 14.7. The optical depths in the upper convection zones for which $\pi F_c / \pi F_r \approx 0.01$ are shown as a function of the effective temperatures for different gravities $\log g$.

If convection transports a major fraction of energy, then there is less energy to be transported by radiation, as is apparent from the energy equation (14.50). Since the temperature gradient is proportional to the radiative flux (see equation 14.10), the temperature gradient is then smaller than in radiative equilibrium (see Fig. 14.8). The total energy transport becomes easier; therefore, a smaller temperature gradient is sufficient in order to transport the required flux.

We can easily convince ourselves that the actual temperature gradient can never decrease below the adiabatic one, because then convection would stop and all the energy would have to be transported again by radiation, which would require a steep temperature gradient. The temperature gradient adjusts in such a way that the stratification remains unstable to convection and that convection transports enough energy to make up for the reduced radiative flux. In Volume 3 we shall discuss how the temperature stratification can be calculated.

So far we have also neglected the radiative energy exchange between the rising or falling gas and the surroundings which reduces the temperature differences and the velocities. Such effects will also be taken into account, when numerical values are calculated in Volume 3.

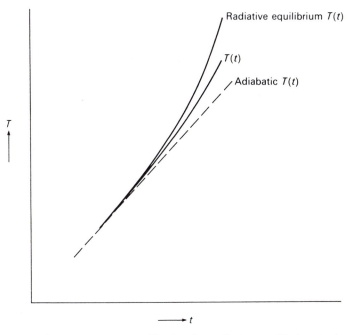

*Fig. 14.8.* The temperature stratification for radiative equilibrium is shown schematically. If the layer becomes convectively unstable and convection contributes to the energy transport, the temperature gradient becomes smaller but always remains larger than the adiabatic gradient.

# 5

# Stellar chromospheres, transition layers, and coronae

## 15.1    Solar observations

We have all seen pleasant pictures of the solar corona taken during solar eclipses when the corona becomes visible. The total light output from the corona is only $10^{-6}$ times the light output from the solar disc. Because the light from the solar photosphere is scattered in the Earth's atmosphere and makes the sky look bright – brighter than the solar corona – we can

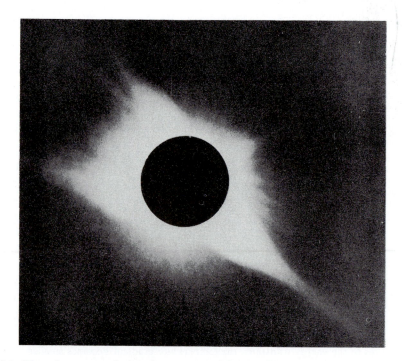

*Fig. 15.1.* The solar corona during the sunspot minimum. (From Kuiper, 1953.)

190

only see the corona when the solar disc is occulted by the moon. On high mountains, where the scattered light from the atmosphere is reduced (when seen from an airplane the sky is much darker) we can also observe the corona by means of the so-called coronagraph, which is a telescope with a built-in disc to occult the light from the solar photosphere.

The solar corona actually looks quite different at different times (see Figs. 15.1 and 15.2). During sunspot maximum the corona looks spherically symmetric, while during sunspot minimum the corona is very extended above the equatorial belt, but shrinks in the other direction. It then looks quite oblate, with a lot of structure in it – bright rays or loops stick out in the equatorial regions. These bright structures look especially impressive on a solar image taken with an X-ray sensitive camera (see Fig. 15.3). These X-ray bright structures are also very prominent on the disc because the solar photosphere does not emit X-rays; only the corona does. Only the corona is hot enough to emit such energetic rays as the X-rays, which have wavelengths of a few angstroms to a few hundred.

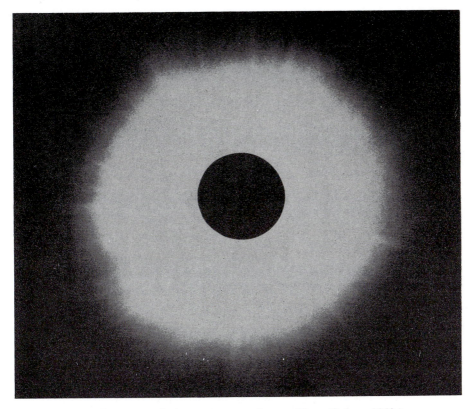

*Fig. 15.2.* The solar corona during sunspot maximum. (From Kuiper, 1953.)

When the wavelengths for the coronal spectral lines were first measured, there were fairly strong lines at 5303, 5694, and 6374 Å. Their identification was completely obscure. Only after B. Edlén in Lund had measured in the laboratory wavelengths of ultraviolet lines of highly ionized lines like Fe IX, which has lost eight electrons, and Fe X, which has lost nine electrons, could he construct energy level diagrams for these ions. With these energy level diagrams, W. Grotrian in Potsdam succeeded in 1939 in identifying some of the coronal lines as being due to very highly ionized ions. Following this discovery, B. Edlén in Lund was able to identify the remaining coronal lines in the visible region as being due to other very highly ionized particles. The line at 5303 Å is due to an iron ion which has lost 13 electrons. The 6374 Å line is due to an Fe ion which has lost 9 electrons, and the 5694 Å line is due to a Ca ion which has lost 14 electrons. For some of these ions, energies of the order of 300 eV are needed to separate the last electron from the next lower ionized particle. The radiation from the solar photosphere certainly does not have such high energy photons. The only way these particles can be ionized to such a high degree is by means of

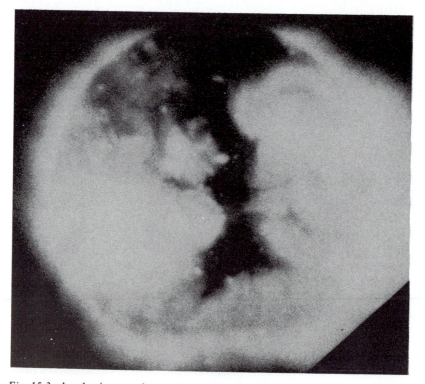

Fig. 15.3. A solar image taken with an X-ray sensitive camera. (From NASA, courtesy Golub and Harvard Smithsonian Observatory.)

collisions with very energetic particles. But to make particles with such high kinetic energies the temperature in the corona has to be one or two million degrees. The X-ray emission of the corona indicates similar temperatures.

There had actually been earlier studies, apparently by several astronomers, investigating the line widths of the coronal lines, which are quite broad. When these widths were interpreted as being due to thermal Doppler broadening, they also indicated temperatures around two million degrees. Before Grotrian's discovery this seemed impossible. Therefore, these studies were not accepted at the time and only became known after Grotrian's and Edlén's coronal line identifications.

Even after those discoveries, it remained a puzzle how the corona could be so hot. We have always emphasized that for the heat flow to go from the inside out, the temperature has to decrease outwards. Now we find this large temperature *increase* outwards. How can the radiation from the relatively cool photosphere with 5800 K flow outwards to and through the hot corona such that it reaches the Earth? Does it not contradict the second law of thermodynamics when the cool photosphere heats the corona up to a million degrees? Or what else is heating the corona? What keeps the photosphere cool between the hot solar interior and the hot corona? All these questions were hard to answer. We are now close to the answers to many of the questions, but before we talk about the energy equilibrium and the heating mechanism for the corona we need to discuss the observations concerning the layers which are below the corona but above the photosphere. Clearly, the temperature cannot jump abruptly from 5800 K in the photosphere to two million degrees in the corona. There must be regions with intermediate temperatures. These are called the chromosphere and the transition layer.

### 15.1.1 *The solar chromosphere*

The chromosphere got its name from the bright colors seen from this layer, which is immediately above the photosphere and in the visual light can only be observed during total solar eclipses a few seconds after second contact and a few seconds before third contact (see Fig. 15.4). The chromosphere has a height of only a few thousand kilometres. Since the spectrum of the chromosphere is seen only for such a short time, it is also called the 'flash' spectrum.

In Fig. 15.5 we show a flash spectrum. It shows strong emission in the hydrogen Balmer lines, but also emission lines of Fe II, Chr II, Si II, etc.

All these are lines which originate in regions with temperatures between 6000 and 10 000 K. Why does the flash spectrum show emission lines? Because we see an optically thin layer (optically thin in the continuum) of hot gas against a background of cool interstellar material (see Fig. 15.5).

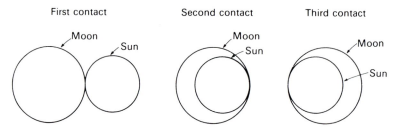

Fig. 15.4 explains the meaning of second and third contact during total solar eclipses. The angular size of the moon is exaggerated relative to the sun.

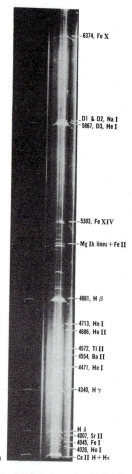

Fig. 15.5. A chromospheric flash spectrum. Several line identifications are given. The coronal lines are seen as whole circles, chromospheric lines as part of a circle. (From Zirin, 1966.)

What about the hot corona you might ask, which is in front as well as behind the chromospheric layer along the line of sight? The corona is so optically thin, which means $\tau_\lambda$ is so small in almost all of the wavelengths, that the intensity emitted from the corona is much smaller than the one emitted from the chromosphere. Therefore, we do not see it in the flash spectrum, except in a few strong coronal lines seen in Fig. 15.5. The chromospheric gas does indeed absorb some of the coronal light, but again, the optical depth of the chromosphere is so small and the coronal intensity is so small that we cannot see this absorption. Only in the chromospheric lines is the optical depth large enough to cause any measurable absorption, but since there is so little coronal light available to be absorbed, the emission

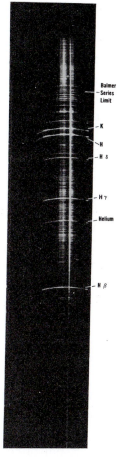

(b)

*Fig. 15.6.* For the solar chromosphere we see an emission line spectrum because we see an optically thin (in the continuum) hot gas against the background of the cool interstellar medium. (From Zirin, 1966.)

in the chromosphere is much stronger than the absorption of the coronal light such that we do indeed see only emission lines.

In the far ultraviolet ($\lambda < 1700$ Å) the chromospheric emission lines can be observed outside of solar eclipses. Of course, we cannot observe in these wavelengths with ground-based observatories because UV light cannot penetrate the Earth's atmosphere. The first UV observations were all done by means of rockets. Now most of these observations are done by satellites. In Fig. 4.12 we show an early ultraviolet spectrum of the sun. For $\lambda > 1700$ Å, an absorption line spectrum from the photosphere is seen. At these wavelengths we see fairly deep layers of the solar atmosphere. For $\lambda < 1700$ Å, the continuous absorption coefficient increases steeply, mainly due to the sharp increase of the Si I absorption coefficient, as we discussed in Section 7.6. For shorter wavelengths therefore, we only see high layers in the atmosphere. In fact, we see the outer layers where the temperature increases; thus we see emission lines, even when we observe the solar disc.

Some people like to say that for these short wavelengths the photospheric light has decreased so much because of the low photospheric temperatures; that we only see the chromospheric emission lines. This, of course, also means that we see the chromospheric emission lines because the chromosphere is hotter than the photospheric layers which we see at these wavelengths.

### 15.1.2  The transition layer

The spectral lines attributed to the chromosphere are lines which are formed in layers with temperatures up to about 15 000 K, except for the helium lines, which are formed in layers with $T \approx 20\,000$ K. Since the temperature must keep rising until it reaches a million degrees in the corona, there must be a layer with temperatures between 20 000 and $10^6$ K. This layer is called the transition layer between chromosphere and corona. It is, of course, only a matter of definition where you think the chromosphere stops and the transition layer begins.

In the spectrum of the transition layer we see lines from different kinds of more or less highly ionized particles.

Generally, we can say that for higher ionization energies the strongest spectral lines of the ions are seen at shorter wavelengths, as can easily be inferred from their energy level diagrams. For higher degrees of ionization the energy level diagrams become more hydrogen-like, but of course all the energies are scaled according to the higher ionization energy, meaning that all the transition energies are also scaled according to the ionization energies, which means the spectral lines all have shorter wavelengths. For He II, for instance, which has an ionization energy four times as large as that of hydrogen, the He II Lyman $\alpha$ line has a wavelength which is one-quarter of that of the hydrogen Ly $\alpha$ line, namely 304 Å. For the ions which we find in the transition zone, which all have higher ionization energies than hydrogen, the strong lines are therefore all in the wavelength region with $\lambda < 2000$ Å. They can only be observed by means of satellite or rocket telescopes. In the solar ultraviolet spectrum seen in Fig. 4.12 lines of transition layer ions like C IV and Si IV are seen.

## 15.2  Stellar observations

We have just seen that now we can observe solar chromospheric and transition layer spectral lines in the ultraviolet without having to wait for a solar eclipse. The ultraviolet observations have also opened up the possibility of observing chromospheric and transition layer spectra of stars similar to the sun. The International Ultraviolet Explorer Satellite (IUE) has been especially successful in this respect. Chromospheric and transition layer spectra of many F, G, and K stars have been observed. In Fig. 15.7 we reproduce the observed spectrum for the bright giant $\beta$ Draconis. Line identifications are given. Most of the lines are attributable to the transition zone. Chromospheric emission is observed in the O I line at 1304 Å and in the C I line at 1657 Å. It is also seen in the central emission in the Mg II lines at 2800 Å (see Fig. 15.8a and b). These lines are especially interesting

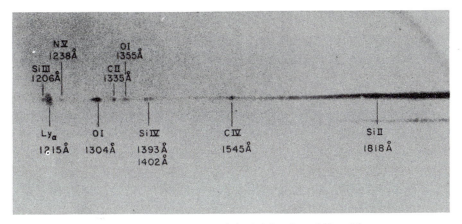

*Fig. 15.7.* The ultraviolet IUE spectrum of the bright giant $\beta$ Dra. The stronger lines are identified with the elements and ionization states which give rise to them. For example, O I is neutral oxygen and Si III is doubly ionized silicon. (From Böhm-Vitense, 1979.)

(a)

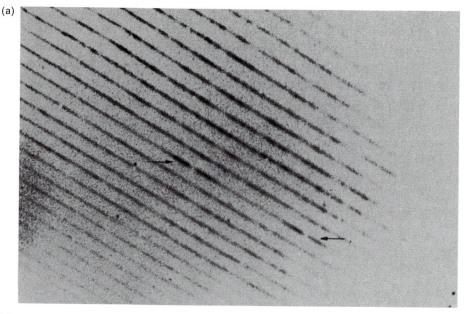

(b)

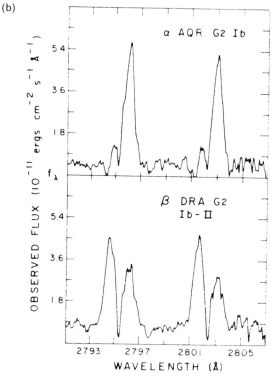

*Fig. 15.8.* (a) An enlarged fraction of an echelle spectrum of β Draconis. The emission cores of the ionized magnesium lines – indicated by arrows – are clearly visible. (Bear in mind that *darker* lines mean more emission from the star.) The $Mg^+$ lines of β Dra show a central emission which is due to the emission in its

because we clearly see the photospheric spectrum with the strong and very broad absorption lines, broadened by collisional broadening. Super-imposed on this photospheric spectrum is the chromospheric emission line which for this star has an intensity which is even higher than the photospheric continuum intensity. For many other stars the chromospheric emission line becomes visible only because the photospheric intensity in the center of the absorption line is so small.

Looking at all the observed stars which show chromospheric or transition layer emission lines and looking at their position in the HR diagram we see that all stars to the red of the Cepheid instability strip show chromospheric emission (see Fig. 15.9). No chromospheric or transition

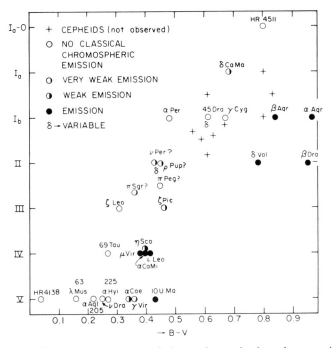

Fig. 15.9. Shown is the position of observed stars in the color magnitude diagram. Stars with chromospheric emission are shown as filled symbols, stars without are shown with open symbols. Half-filled symbols indicate uncertain cases. The Cepheid instability strip is plotted. All filled symbols are on the cool side of the instability strip. (From Böhm-Vitense and Dettmann, 1980.)

*Fig. 15.8 continued*
chromosphere. (From Böhm-Vitense, 1979.) (b) The intensity tracings in the centers of the $Mg^+$ lines at 2795.5 Å and 2802.7 Å for stars $\alpha$ Aquarii and $\beta$ Draconis. The chromospheric emission with the central interstellar absorption is clearly seen. (From Stencel *et al.*, 1980.)

layer emission lines are seen for stars on the blue side. Cepheids may show chromospheric or transition layer emission lines at some phases of their pulsational cycle.

The Cepheid instability strip also marks the boundary line for stars with a hydrogen convection zone. For stars on the blue side of the Cepheid instability strip the hydrogen ionization occurs so high in the photosphere that convection becomes very inefficient. Convective velocities are of the order of cm/sec. We thus suspect that the heating of the outer layers of these F, G, and K stars is related to the existence of the hydrogen convection zone.

Another observation seems to be important; namely, for cool luminous stars no transition layer emission lines are observed (see Fig. 15.10). If they are there, the intensities are below the limit of detectability. It is interesting to note that for these stars rather strong stellar winds are observed, the origin of which is not yet understood.

Another interesting observation is the fact that for stars of spectral type later than F5 the chromospheric emission seems to decrease with increasing

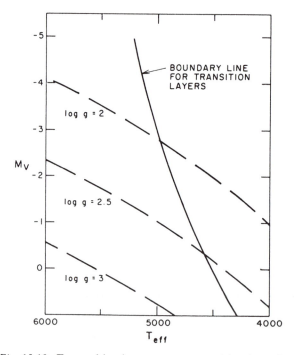

*Fig. 15.10.* For cool luminous stars no transition layer lines are observed as first discovered by Linsky and Haisch, 1979. The Linsky Haisch boundary line for transition layer emission is sketched in this $T_{\text{eff}}$ magnitude diagram. Lines of constant gravity are shown as dashed curves. (From Böhm-Vitense, 1986.)

age of the star, as do the rotational velocities. In Fig. 15.11 we show the diagram plotted by Skumanich (1972), who first pointed out this correlation for the Ca II chromospheric emission, seen in the centers of the broad photospheric absorption, and the ages of the stars. The Ca II chromospheric emission lines are similar to the Mg II emission lines but much weaker. They are at wavelengths of 3968 and 3933 Å, and can therefore be observed from ground-based observatories. They were studied long before ultraviolet observations could be made.

Skumanich also put together data for rotational velocities of the stars of different ages. Both chromospheric emission and rotational velocities decrease with increasing age of the stars.

For the determination of the small rotational velocities of the late F and G stars, it is to our advantage that the Ca II chromospheric emission lines of the *sun* vary in intensity with the sunspot cycle. They are stronger during sunspot maximum. They are also stronger if large sunspot groups are seen on the visible hemisphere. If such spot groups persist for more than one solar rotation period – which is 28 days – the Ca II emission line strengths show a variation with a period of 28 days. Similar periods of variation can be found for other solar type stars and are believed to be also due to the

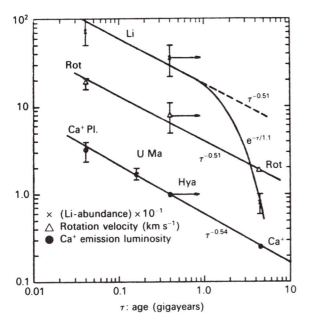

Fig. 15.11. The Ca$^+$ chromospheric emission line strengths are plotted as a function of age of the stars. Also shown are the average rotational velocities of stars of different ages. (According to Skumanich, 1972.)

rotation period. In this way rather small rotational velocities can be determined. The correlation between rotation and chromospheric emission can thus be checked.

Coronal X-ray emission has also been observed for many stars by means of the Einstein satellite. X-rays have been seen for nearly all kinds of late type stars which show transition layer emission lines. X-rays have been observed, however, for all types of hot luminous stars also, i.e., for O and B type stars. The origin for these latter X-rays is at present not understood. It seems unlikely that these stars have a solar-type corona.

## 15.3    Theory of stellar chromospheres, transition layers and coronae

### 15.3.1    *The energy loss mechanisms*

For a temperature *equilibrium* in the outer layers of the stars we must again require that in any given layer there is as much energy taken out as there is energy fed into this layer. If there were only radiative energy taken out and only radiative energy put in, then we would have radiative equilibrium. We know that in this case in a grey atmosphere the surface temperature would be given by (see equation 6.38)

$$T^4(\tau \approx 0) = \tfrac{3}{4}T^4_{\text{eff}} \times \tfrac{2}{3} = \tfrac{1}{2}T^4_{\text{eff}}.$$    15.1

For a non-grey atmosphere we expect an even lower temperature at the surface. This means that the amount of radiative energy coming from the photosphere provides only enough energy to keep the chromosphere at this low temperature. As we observe the chromospheres to be hotter, there must be an additional energy source. This cannot be a thermal energy source, because the second law of thermodynamics tells us that the cool photosphere can never heat the chromosphere to higher temperatures than the photospheric temperature by any thermal process. There are no such restrictions with respect to heating by mechanical or electrical energy (you can make a fire by hitting one cool rock with another cool rock). We must therefore conclude that some sort of mechanical or electrical energy is heating the outer layers of the stars with chromospheres. We observe large velocities in the solar granulation (see Volume 1, Chapter 18) which appear to be turbulent. Lighthill (1951) and Proudman (1952) calculated that turbulent velocities create sound waves (the jets from a plane create a lot of noise) and that the energy in these sound waves, or acoustic waves, increases with the eighth power of the turbulent velocity $v_{\text{turb}}$. It is also proportional to the density in the turbulent layer. This means the acoustic flux $F_{\text{ac}}$ is given by

$$F_{ac} \propto \rho v_{turb}^8 / c_s^5, \qquad\qquad 15.2$$

where $c_s$ is the speed of sound. These acoustic waves travel upwards from the photosphere and enter layers of lower density, the energy flux being given by $\frac{1}{2}\rho v^2 c_s$ (energy density $= \frac{1}{2}\rho v^2$, where $v$ is the particle velocity, multiplied by propagation speed $c_s$). If they do not lose energy their velocity must increase because of the decreasing density. This means they steepen up to shock waves. While the acoustic waves as such do not lose much energy once they become shock waves they are damped very rapidly, which means they transfer a lot of energy to their surroundings, as we know from experience. This energy transfer to the surroundings, of course, means that the surroundings experience an energy input, by which it gets heated.

In the presence of magnetic fields these sound waves and shock waves are modified, becoming so-called magnetohydrodynamic waves (MHD waves) of different kinds, but there is always one kind which behaves similarly to sound waves and which can form shock waves. How quickly are these acoustic waves damped? The answer depends on the period of the waves. In the lower chromosphere we observe oscillations with periods of about 300 seconds. If this number is used one finds that in the solar chromosphere with a temperature of about 10 000 K the shock waves are dissipating energy with a damping length $\lambda$ of about 1500 km. This means that the change of the shock wave flux, $F_{mech}$, with height is given by

$$\frac{dF_{mech}}{dh} = -\frac{1}{\lambda} F_{mech}, \qquad \text{where } \lambda \approx 1500 \text{ km.} \qquad 15.3$$

This damping length is comparable to the height of the solar chromosphere. The energy amount $F_{mech}/\lambda$ is the amount dissipated into heat in each layer and which heats that layer of the chromosphere. Integration of equations (15.3) yields

$$F_{mech} = F_{mo} e^{-h/\lambda}, \qquad\qquad 15.4$$

where $F_{mo}$ is the mechanical flux at $h = 0$. This means the amount of energy deposited in layers with increasing height is decreasing. Nevertheless, the higher chromospheric layers have a higher temperature than the lower layers. In order to understand this we have to look at the energy balance. But first we shall discuss another heating mechanism.

### 15.3.2 *The energy input by Joule heating*

We observe a change in the chromospheric emission during periods with large sunspots and large solar activity. Since we know that these activities are related to the magnetic field which we observe in the sunspots, we must consider magnetohydrodynamic waves also. While there are always waves which are similar to acoustic waves, there are also always waves which are similar to the Alfvén waves, named after the physicist who first discovered their existence. For the existence of these waves the changes in the magnetic field, also caused by the turbulent motions in the convection zone, are very important. Changes in the magnetic fields always cause electric currents, as we well know from experiments. If electric currents flow through a medium with some resistivity they will heat that medium. This heat is called the Joule heat. The stellar material in the chromosphere is a very good conductor due to all the free electrons, especially in the low density chromosphere and transition layer, where there are few collisions. Therefore, not much Joule heat is generated. The energy loss of the Alfvén waves and the magnetohydrodynamic waves, which are similar to the Alfvén waves, is small and the damping length $\lambda_A$ is large, but we do not know offhand which energy flux is larger – that of the shock waves or that of the Alfvén waves. Therefore, we have to keep in mind both types of waves for the energy input into the outer layers of the sun and similar stars.

Alfvén waves are also damped by other mechanisms, which may be even more important, but it may suffice here to mention just one.

### 15.3.3  *The energy loss by radiation*

If there were only energy input and no energy loss, the temperature would keep increasing all the time. Since this is not the case in the chromospheres, transition layers and coronae, there must also be an energy loss which balances the energy input.

As the chromospheres, transition layers, and coronae can be seen, they must radiate, which means they lose energy due to the radiation. We want to determine these energy losses. In principle, we can measure the amount of radiation emitted, which gives us the energy loss. In practice, there are some difficulties because there may be a fair amount of radiation in a wavelength region which we cannot observe except for the sun. These are the wavelengths shortward of 912 Å, in which the interstellar hydrogen absorbs all the light for most stars. The sun, a few nearby white dwarfs and possibly a few main sequence stars are near enough such that this does not happen. The solar chromosphere loses a considerable amount of radiation in that spectral region. We do not just want to know how much energy a given star loses, but, in order to understand the energy balance, we want to know how this energy loss depends on the temperature and the density.

The emission in the Lyman continuum is due to free-bound transition, which means due to a capture of an electron, which means a recombination of an electron and a proton to form a neutral hydrogen atom in the ground level (see Chapter 7). The radiation in the Balmer continuum is due to a recombination into the second quantum level. This electron in the second quantum level will subsequently fall into the ground level and in the process emit a photon in the Ly α line. The energy emitted is therefore determined by the number of recombinations, which requires that the electron and proton come very close together; in other words, to the number of collisions between a proton and an electron. The number of such collisions increases proportional to the number of electrons $n_e$ per cm³ and also proportional to the number of protons $n_p$ per cm³, i.e.,

$$\text{number of collisions} \propto n_e n_p. \qquad 15.5$$

Additional energy loss is caused by collisional excitations due to collisions with electrons, and subsequent emission of a photon. The number of recombination or excitation collisions depends also on the velocity of the particles which in turn depends on the temperature. In the chromosphere most of the electrons originate from the ionization of hydrogen. Therefore, we can generally say that $n_p \approx n_e$. The amount of radiated energy $E_{rad}$ is then proportional to the number of collisions. We find for hydrogen recombination

$$E_{rad} = n_e^2 f_{H,r}(T). \qquad 15.6a$$

For collisional excitations and subsequent radiation we find for hydrogen lines

$$E_{rad} = n_e n_H \cdot f_{H,L}(T), \qquad 15.6b$$

where $f_H(T)$ depends on the collisional excitation rates.

So far we have only considered the energy loss in the hydrogen continua and lines. Similar arguments can be made for other important lines, if the upper levels are mainly populated by collisions with electrons. If we are talking about magnesium lines, then the number of $Mg^+$ ions is, of course, important; or, for the $Ca^+$ lines, the number of $Ca^+$ ions is important. Given a chemical composition, the number of collisions for the different ions can always be calculated and is always proportional to the square of the density. For a given chemical composition the radiative energy losses can therefore always be written in the form of equation (15.6), where $f(T)$ can be calculated for each chemical composition.

Clearly, not every collision causes a recombination, that is, a capture of the electron. Therefore, we also have to determine the probability that a collision leads to a recombination or to an excitation of a bound electron into a higher energy level. These probabilities can either be calculated – but the calculations are not very accurate for complicated ions – or they can be measured in the laboratory. For many ions the probabilities for collisional recombinations or excitations, also called the collisional cross-sections, are not yet known exactly. Because of this the functions $f(T)$ calculated by different authors come out somewhat different. In Fig. 15.12 we show the results of several authors. They all agree on the fact that $f(T)$ increases rather steeply with increasing temperature up to temperatures $T \approx 15\,000$ K. At this temperature hydrogen is nearly completely ionized and no longer contributes much to the energy loss. Therefore, $f(T)$ decreases slightly at this temperature, until helium starts to come into play and also other heavier ions. For $T > 20\,000$ K the $f(T)$ starts increasing again. For temperatures $T > 1.6 \times 10^5$ K most of the abundant elements are almost completely ionized. At this temperature the radiative losses do finally decrease with increasing temperature.

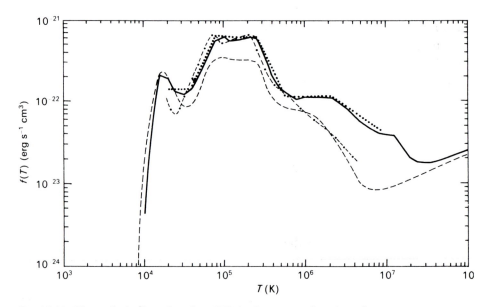

Fig. 15.12. The radiative loss function $f(T)$ is shown as a function of temperature. The different curves shown were calculated by different authors (from Rosner, Tucker and Vaiana, 1978). Shown are the results of Pottasch (1965) [$\cdot$---$\cdot$], McWhirter *et al.* (1975) [---], Raymond (1978) [———], and an analytic fit to Raymond's curve [$\cdots\cdots$].

This temperature dependence of the radiative energy losses is very important for the understanding of the temperature stratification in the outer layers of solar type stars.

### 15.3.4  *Conductive heat transport*

In high temperature regions the particles have high velocities. They may travel into adjacent lower temperature regions, and thereby carry their high kinetic energy into the cooler region. At the same time the particles from the lower temperature region with their lower velocities travel into the high temperature region, but since they have lower velocities they carry only a small amount of kinetic energy into the high temperature region. There is therefore a net energy transport from the hot into the cool region. This is what we call heat conduction. If the density is low each particle experiences only few collisions and can travel further, thereby transporting the energy further. This makes its heat conduction more efficient. (In white dwarfs they can travel very far, in spite of the high densities.) On the other hand, for low densities there are fewer particles which can contribute to the heat conduction. These two effects cancel and make the heat conduction in transition layers and coronae independent of the density. However, with increasing velocity of the particles collisions become less frequent, which means the collisional cross-sections become smaller. Therefore, the heat conduction increases with increasing temperature. One finds that in completely ionized gas the heat flux due to heat conduction $F_c$ is given by

$$F_c = -\eta T^{5/2} \frac{dT}{dh}, \qquad\qquad 15.7$$

where $\eta$ is about $10^{-6}$ cgs units.

The minus sign says that the heat flux goes towards smaller temperatures. Generally in stellar atmospheres conductive heat transport is unimportant, except in white dwarfs and in stellar transition layers and coronae. Why? The conductive heat flux carries energy from the transition layer and corona back down to the chromosphere and photosphere. If through the transition layer the conductive flux remains constant, then no heat energy is taken out or put in, just the same as in radiative equilibrium, where no energy is deposited when the radiative flux remains constant. If, however, the conductive flux increases downwards, then this additional energy must be taken out of the layer where the flux increases. An increasing conductive flux downward then constitutes an energy loss to that layer. If the downward flux becomes smaller, this requires an energy input into the layer where the flux decreases. (We always mean here the absolute value of the flux. We have to be careful, since in the mathematical sense the conductive flux is negative because it goes to decreasing heights.)

### 15.3.5 *The energy balance in the chromospheres, transition layers and coronae*

(i) *The lower transition layer.* We discuss the lower part of the transition layer first, because it is the easiest to understand. From Fig. 15.12 it is apparent that for temperatures between 30 000 K and about $10^5$ K the radiative loss function $f(T)$ can be represented by a simple power law. In this region we have approximately

$$f(T) = 10^{-22} \times (T/30\,000)^\beta, \qquad\qquad 15.8$$

where $\beta$ is around 2.0.

The value of $\beta$ is not very well determined. For a given gas pressure $P_g = 2P_e$, where $P_e$ is the electron pressure. The electron density $n_e$ is given by

$$n_e = \frac{P_g}{2kT}. \qquad\qquad 15.9$$

For a given gas pressure in the transition layer we thus obtain for the radiative energy losses per cm$^3$ s

$$E_{\text{rad}} = f(T)\frac{P_g^2}{(2kT)^2}. \qquad\qquad 15.10$$

In Fig. 15.13 we have plotted $E_{\text{rad}}$ for different values of the gas pressure between $\log P_g = -0.8$, and $\log P_g = -0.95$. Suppose now that we have a given mechanical energy input $E_{\text{in}} = F_m/\lambda$. If $E_{\text{in}} = E_{\text{rad}}$, as required by the energy balance if no other energy loss mechanism is working, then we see immediately from Fig. 15.13 that for the lower gas pressures we need a higher temperature in order to achieve this balance. The higher $f(T)$ has to make up for the lower $n_e^2$. Because of the hydrostatic equilibrium the gas pressure in any atmosphere has to decrease for increasing height. Therefore, the temperature increases with increasing height. Even for a height independent energy input the temperature has to increase for lower gas pressure, i.e., for increasing height. Even if the energy input decreases somewhat with increasing height, the temperature still has to increase, as demonstrated in Fig. 15.14. There is, however, a limit to the decrease of energy input for which we still find increasing temperatures for decreasing energy input. If the energy input decreases faster than $P_g^2$ then the temperature does not increase any more with height, as can be seen from

Fig. 15.15. The gas pressure decreases with a scale height $H$, as we have discussed in Chapter 9. The $P_g^2$ therefore decreases as

$$P_g^2 = P_{g_0}^2 e^{-2h/H},\qquad 15.11$$

which means it decreases with half the scale height as the gas pressure

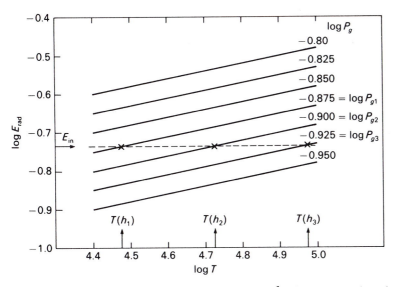

Fig. 15.13. The radiative energy losses in erg per cm³. The $E_{rad}$ are plotted as a function of temperature for seven different values of the gas pressure. For a given energy input $E_{in}$ in erg per cm³, higher equilibrium temperatures $T_e$ are needed for lower gas pressures in order to balance the energy input $E_{in}$ by the radiative energy losses.

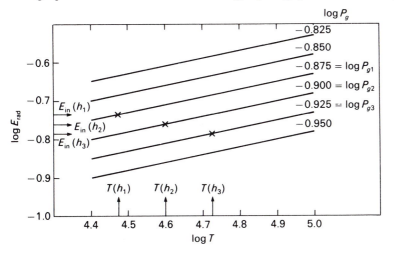

Fig. 15.14. As Fig. 15.13, but the energy input $E_{in}$ decreases with increasing height. Even for an energy input which is decreasing with height, the temperature still has to increase if the decrease of $E_{in}$ is not too large.

itself. The energy input decreases with the damping length (or scale height) $\lambda$. We therefore must conclude that the temperature does not increase outward if $\lambda$ becomes less than half the pressure scale height, i.e., if

$$\lambda = H/2 \quad \text{where } H = \frac{R_g T}{\mu g_{\text{eff}}}. \qquad 15.12$$

(For the definition of $g_{\text{eff}}$ see equation 9.45.)

Since we have not yet determined the energy input mechanism, we cannot yet say how large $\lambda$ is, but we know that the pressure scale height increases with decreasing gravitational acceleration. $H$ is therefore much larger for the luminous stars. Unless $\lambda$ grows just as fast with increasing luminosity as the pressure scale height, we may expect that for very low gravity stars, i.e., cool luminous stars, the damping length $\lambda$ becomes smaller than half the pressure scale height at some point in the chromosphere. From that point, the temperature will not increase outwards. We think that this is the reason why cool luminous stars do not have transition layers and coronae.

Making use of the energy equilibrium equation and approximating $f(T)$ by $f(T) = BT^\beta$

$$\frac{F_m}{\lambda} = f(T)n_e^2 = BT^\beta n_e^2 = BT^{\beta-2}(P_g/2k)^2, \qquad 15.13$$

we can calculate the temperature stratification in the lower transition zone. If $\lambda \gg H$ we can assume that $F_m/\lambda$ is approximately constant in the lower

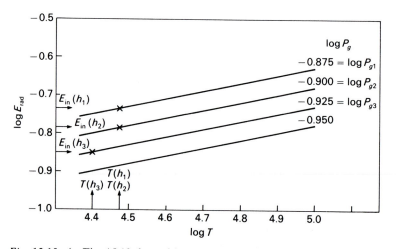

Fig. 15.15. As Fig. 15.13, but with an energy input $E_{\text{in}}$ which is strongly decreasing with height. If $E_{\text{in}}$ decreases faster than $n_e^2$, then the equilibrium temperature does not increase with height. It may decrease.

part of the transition region, which means according to equation (15.13) that

$$\frac{F_m}{\lambda} = \frac{B}{(2k)^2} T^{\beta-2} P_g^2 = \text{constant},$$ 15.14

or

$$(\beta - 2)\frac{d \ln T}{dh} + 2\frac{d \ln P_g}{dh} = 0,$$ 15.15

$$\frac{d \ln T}{dh} = -\frac{2}{(\beta-2)}\frac{d \ln P_g}{dh} = \frac{2}{(\beta-2)}\frac{\mu g_{\text{eff}}}{R_g T} = \frac{A}{T},$$ 15.16

or

$$\frac{dT}{dh} = A = \frac{2}{(\beta-2)}\frac{\mu g_{\text{eff}}}{R_g}.$$ 15.17

The temperature gradient is constant, the temperature increases linearly with height. This is a very simple relation which matches observations very well. With this simplified approximation the equilibrium temperature would decrease outwards if $\beta$ is less than 2.

If $F_m/\lambda$ is not constant with height, but

$$F_m = F_{mo} \exp\left(-\int_0^h \frac{1}{\lambda} dh\right) \quad \text{and} \quad \frac{d \ln F_m}{dh} = \lambda$$

and if, for instance, $\lambda = \lambda_0 T^\alpha$, i.e., if $\lambda$ increases with increasing height, then we derive from equations (15.14) and (15.13)

$$T^{\beta+\alpha-2} = \frac{(2k)^2}{B}\frac{1}{P_g^2}\frac{F_m}{\lambda_0}.$$ 15.18

Logarithmic differentiation yields

$$(\beta+\alpha-2)\frac{d \ln T}{dh} = -\frac{2 d \ln P_g}{dh} + \frac{d \ln F_m}{dh}$$

Making use of equations (15.3), (9.3) and (9.4) we then find

$$\frac{d \ln T}{dh} = \frac{2}{(\beta+\alpha-2)}\frac{\mu g_{\text{eff}}}{R_g T}\left[1 - \frac{H}{2\lambda}\right].$$ 15.19

As we saw when discussing the radiative energy losses, the emission line intensities are also proportional to the electron densities $n_e^2$. The values for the electron densities can therefore be obtained from the measured emission line intensities. Knowing $n_e^2$ we can determine the ratio $F_m/\lambda$ from the left-hand side of equation (15.13).

(*ii*) *The upper part of the transition layer.* Fig. 15.12 shows that for temperatures larger than about $10^5$ K the ratiative loss function $f(T)$ no longer increases with increasing temperature, which means $\beta \leqslant 0$. With decreasing density the radiative losses therefore decrease; an increase in temperature cannot prevent this. If the energy input is larger than the radiative losses for $T > 10^5$ K, then the transition layer cannot get rid of it by means of radiation. There is then a surplus of energy input; the atmospheric layer must heat up. It continues to heat up until some other energy loss mechanism becomes effective. Due to the heating, a very steep temperature gradient develops at this temperature which leads to a relatively large conductive heat flux down into the lower layers. If the temperature gradient is large enough this conductive heat flux is able to take out the surplus energy. There can be an energy equilibrium between the energy input $dF_m/dh$ and $dF_c/dh$, where $F_c$ is the conductive heat flux. There is conductive flux coming down from the corona. At each layer in the transition zone there is some mechanical energy input and some radiative energy losses. If the energy input is larger than the losses, then the net energy input must also be transported down by the conductive flux. This means at each layer some energy is added to the conductive flux, which keeps increasing on its way down (see Fig. 15.16). The largest conductive flux is then found at the bottom of this upper part of the transition layer, that is, in the layer with $T \approx 10^5$ K.

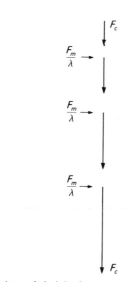

Fig. 15.16. At each height $h$ some energy $E_{in}$ is put in. For an equilibrium this energy has to be added to the conductive flux in order to get rid of this energy and keep the temperature constant.

If there were no energy lost in these upper layers by radiation or by some other means, then all the mechanical flux which entered at the bottom of these high temperature regions must return in the form of conductive flux. In reality, there is also some energy lost due to radiation and a somewhat larger amount is lost due to the stellar wind, as will be discussed in the next chapter. In fact, at each layer in the transition region the amount of conductive flux coming down must equal the amount of mechanical flux going up, minus the energy loss due to the wind and the radiation in the layers above the one under consideration. As the stellar wind takes out energy mainly at the top of the corona and the radiative losses are small, we can say that at each layer

$$|F_c(h)| = F_m(h) - \Delta F_{\text{wind}} - \Delta F_{\text{rad}}(h), \qquad 15.20$$

where $\Delta F_{\text{wind}}$ is the energy loss per cm$^2$ s due to the stellar wind, and $\Delta F_{\text{rad}}(h)$ the radiative losses per cm$^2$ s above $h$.

Combining equations (15.7) and (15.20), we derive

$$|F_c| = \eta T^{5/2} \frac{dT}{dh} = \eta \frac{d}{dh} (T^{7/2})\tfrac{2}{7} = F_m(h) - \Delta F_{\text{wind}} - \Delta F_{\text{rad}}$$

$$= F_{m_0} e^{-h/\lambda} - \Delta F_{\text{wind}} - \Delta F_{\text{rad}}(h). \qquad 15.21$$

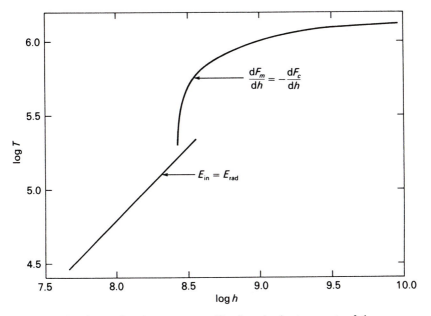

Fig. 15.17. The observed temperature stratifications in the two parts of the transition region of Procyon are shown as a function of height (Jordan *et al.*, 1984). The difference in the temperature gradient between the lower and the upper part of the transition layer is clearly seen, the transition occurring at $T \sim 1.6 \times 10^5$ K.

Here, $F_{wind}$ can be considered to be the same for all heights except in the high corona.

The integration of the right-hand half of this equation between $h = h_0$ and $h$ yields

$$T(h)^{7/2} - T(h_0)^{7/2} = \frac{7}{2}\frac{1}{\eta}\left[F_{mo}\lambda(1-e^{-(h-h_0)/\lambda}) - \Delta F_{wind}(h-h_0) - \int_{h_0}^{h}\Delta F_{rad}(h)\,dh\right].$$

15.22

Here $F_{mo}$ is now the flux at $h = h_0$, and $h_0$ is the height of the layer in which $T = 10^5$ K. The temperature gradient is largest at the bottom, where the mechanical flux and therefore the conductive flux is probably largest, and where the conductivity is the smallest.

At any given point in the atmosphere the temperature is higher, the larger $F_{mo}$ and the larger $\lambda$, the damping length.

In Fig. 15.17 we show as an example the temperature stratification for the transition layer in Procyon, an F5 IV star. This star has a fairly large mechanical energy flux. The different temperature gradients in the lower and the upper part of the transition layer are clearly distinguishable.

# 16

# Stellar winds

## 16.1 Observations of stellar winds

### 16.1.1 The solar wind

The study of the orientation of comet tails revealed that one part of the comet tail always points radially away from the sun, rather than trailing behind the comet. This indicates that some force is acting on the particles in the tail pushing them radially away from the sun. The first suspicion was that solar photons being absorbed and scattered by these particles might transfer enough momentum. As we discussed earlier (see Chapter 9), each absorbed or scattered photon transmits on average a net momentum of $h\nu/c$ to the absorbing particle in the direction of propagation of the photon, which, in this case, means away from the sun (see Fig. 16.1). It turned out, however, that the light pressure force is much too small to explain the radial direction of the comet tails. Biermann (1951) was the

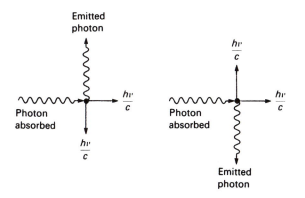

*Fig. 16.1.* Each absorbed photon transmits a momentum $h\nu/c$ to the absorbing particle in the direction of propagation of the photon. The emitted photons transmit a momentum in the direction opposite to the propagation direction. The momenta transmitted by the isotropically re-emitted photons cancel on average. The momenta of the absorbed photons add.

first to suggest that there must be a particle stream with high velocities transferring momentum radially away from the sun to the comet tail particles when they hit them. These high velocity particles moving radially outwards from the sun are called the solar wind.

In the meantime, the densities and velocities of these solar wind particles have been measured by means of rockets and satellites. It has also been detected that they carry magnetic fields with them. These solar wind particles are also the ones which populate the van Allen belts in the Earth's ionosphere. The particle density $n$ in the wind is about five particles per cm$^3$ at the distance of one astronomical unit from the sun. The velocities of the particles depend on the area on the sun from where they started. Particles originating in regions of the so-called coronal holes have velocities up to 700 km s$^{-1}$, while particles originating from the average, so-called quiet solar surface have velocities of the order of 200 to 300 km s$^{-1}$. The total mass flow from the sun can then be estimated to be

$$m = nm_{\mathrm{H}}v4\pi d^2 \qquad \text{where } d = 1\,\text{AU}, \tag{16.1}$$

which turns out to be about $10^{19}$ grams per year, or $10^{-14}\,\mathrm{M}_\odot$ per year. Even during its entire lifetime, which is about $10^{11}$ years, the total mass loss of the sun, if it remains constant, is only 0.1% of its mass.

### 16.1.2  Winds of cool luminous stars

Stellar winds of cool luminous stars were first studied by Deutsch (1956). The amount of mass loss for these cool, luminous stars can be inferred from the line profiles of the Ca$^+$ line at 3933 Å. In these cool, luminous stars the Ca$^+$ lines are not symmetric, but show absorption components on the short wavelength side of the center or show extended wings toward the short wavelength side. In Fig. 16.2 we show an example of such a line profile. The velocities shown by these blue-shifted lines are much smaller than those observed in the solar wind; they are only about 50 km s$^{-1}$. The total number of atoms in the line of sight, also called the column density (see Fig. 16.3), absorbing in these blue-shifted components can be obtained from the line strength and fitting the two Ca$^+$ lines at 3933 Å and 3968 Å on a curve of growth (see Chapter 12). While the velocities measured for the winds of cool, luminous stars are about a factor of 5 to 10 lower than for the solar wind, they refer to layers near the star where the particle densities are still $n \approx 10^8$ per cm$^3$, and the radii of the stars are about a factor of 100 larger than the solar radius. The surface area is therefore larger by a factor of $10^4$. The mass loss for these stars

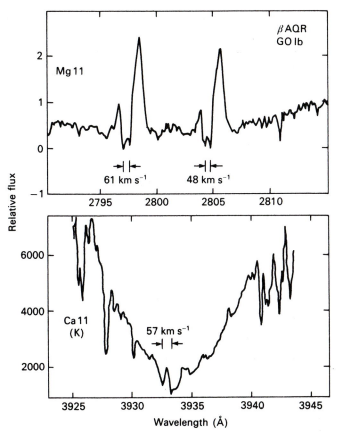

Fig. 16.2. The profile of the Ca$^+$ 3933 Å line in the supergiant $\beta$ Aqr. An absorption component is seen displaced to shorter wavelengths by 0.75 Å, corresponding to an outflow velocity of 57 km s$^{-1}$. Similar displaced components are seen in the Mg$^+$ lines at 2795.5 and 2802.7 Å. The longer wavelength component is due to an interstellar absorption line. (From Hartmann, Dupree and Raymond, 1980.)

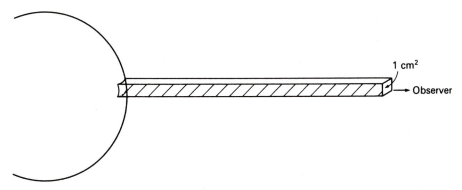

Fig. 16.3. The total number of absorbing atoms or ions in a column of 1 cm$^2$ cross-section extending from the stellar surface to the observer is called the column density.

comes out to be about $10^{-8} M_\odot$ to $10^{-5} M_\odot$ per year. (The exact value is still difficult to determine.) The lifetime for these stars as luminous red giants is estimated to be about $10^5$ to $10^6$ years. They may therefore lose a considerable fraction of their original mass during this stage of their evolution.

### 16.1.3 Stellar winds in hot luminous stars

While some hot luminous stars, such as P Cygni, were for a long time known to have strong mass loss, these kinds of stars were considered to be anomalous objects. Only when the Copernicus satellite was able to take high resolution spectra of many stars in the ultraviolet region, which cannot be observed from ground-based telescopes, was it discovered that strong winds, although less strong than in P Cygni, are a common phenomenon among hot, luminous stars. It was then seen that all such stars show very asymmetric line profiles of the ultraviolet lines of $C^{3+}$ or $N^{4+}$ or $Si^{3+}$, or of several of them. Many such stars show line profiles of these ions, which look similar to the line profiles seen in P Cygni in the optical region. In Fig. 16.4 we show an example of such a P Cygni profile, as all these line profiles are called now.

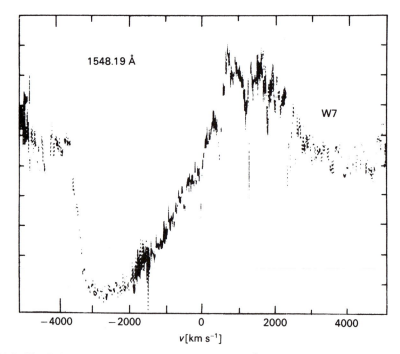

*Fig. 16.4.* The P Cygni profile of the C IV lines at 1550 Å as seen in the spectrum of the O4 V star 9 Sgr.

Such P Cygni line profiles can be interpreted as being caused by a spherically symmetric, expanding shell, as we explain in Figs. 16.5 and 16.6. When the observer is in the direction of the bottom of the page, the material in the shaded area is in front of the star and contributes to an

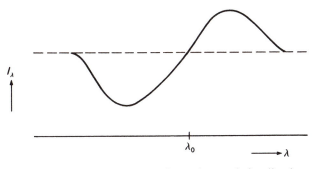

*Fig. 16.5.* A P Cygni profile of an absorption–emission line is expected to be observed for an expanding shell of gas around a star.

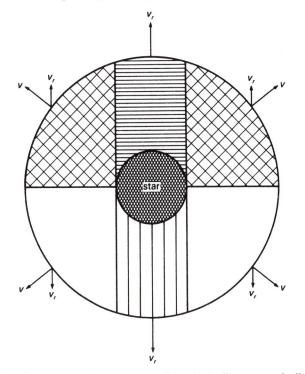

*Fig. 16.6* demonstrates the geometry of a spherically symmetrically expanding shell around a star (dark area). The shaded area in front of the star contributes to the absorption component in the P Cygni profile which is shifted toward shorter wavelengths. The white areas lead to emission at shortward shifted wavelengths. The cross-hatched area leads to emission shifted to longer wavelengths than the laboratory wavelength $\lambda_0$. The horizontally shaded column behind the star is obscured by the star and cannot be seen.

absorption line in the stellar spectrum. All this material is moving towards the observer; therefore, the absorption line is shifted to shorter wavelengths than the rest wavelength $\lambda_0$. The material in the light areas on both sides of the absorbing material also has velocity components toward the observer, but this material is not in front of the star. Therefore, it cannot absorb light coming toward the observer from the star. This material only emits light. It therefore contributes to a slightly blue-shifted emission line. The material moving out perpendicularly to the line of sight of the observer contributes to an unshifted emission line, while the material in the cross-hatched area contributes to an emission line shifted to longer wavelengths because in these regions the velocities have components pointing away from the observer. The material behind the star, which would give an emission line component with the highest redshift, is obscured by the star and does not contribute to the observed spectrum. The largest redshift in the emission line is therefore smaller than the largest blueshift in the absorption line. If we add up all the contributions of the different parts of the shell we obtain a P Cygni profile in which the emission starts at somewhat shorter wavelengths than the rest wavelength, but since the stellar spectrum also has an absorption line, this shell emission still falls in the absorption of the stellar line. Apparently, the emission starts just about at the rest wavelength $\lambda_0$, the unshifted laboratory wavelength. The exact line profile does, of course, depend on the density and velocity distribution within the shell. These are not independent because for a constant mass flow we must have

$$\text{const.} = \rho(r)v(r)4\pi r^2 = \dot{m}. \qquad 16.3$$

This means if $v(r)$ is given, then one value of $\rho(r)$, namely $\rho(r_0)$, completely determines the mass loss rate. Instead, we can also determine the column density of the absorbing gas in front of the star from the strength of the absorption line and in this way determine the mass in the shell which gives us the mass loss. Mass loss rates have been determined in this way for many luminous hot stars. The velocities for the hottest main sequence stars are observed to be as high as 3000 to 4000 km s$^{-1}$. The mass loss rates $\dot{m}$ come out to be about $\sim 10^{-6}$ M$_\odot$ per year. In Volume 3 we shall see that the lifetime of these stars is also of the order of $10^6$ years. These stars may therefore lose a considerable fraction of the original mass during their lifetime on the main sequence, which has considerable influence on their further evolution.

While the general opinion is that the material in the shell is accelerated by the absorption of photons, it is still a puzzle how so much momentum can be transferred to any given particle in the wind. The momentum transfer of each photon is about

$$\Delta p = h\nu/c \sim (6 \times 10^{-27}) \times (3 \times 10^{15})/3 \times 10^{10} = 6 \times 10^{-22} \text{ [cgs]},$$

while the momentum of each wind particle actually is

$$p = m_H \times (3.6 \times 10^8) = 6 \times 10^{-16} \text{ [cgs]}.$$

Each particle therefore has to absorb about $10^6$ photons in order to obtain such high velocities, yet the gas in the wind has to be optically thin except in some lines. A large number of lines must be contributing to the radiative acceleration.

## 16.2 Theory of stellar winds

### 16.2.1 *The impossibility of hydrostatic equilibrium*

It was Parker (1958) who first developed a theory which could qualitatively explain the solar wind.

We have learned that the pressure stratification in a stellar atmosphere is governed by the hydrostatic equilibrium. For an isothermal plane parallel atmosphere with constant gravitational acceleration $g$ the pressure decreases as

$$P_g = P_{g0} e^{-h/H} \qquad \text{with } H = \frac{R_g T}{\mu g}. \qquad 16.4$$

The pressure decreases exponentially but never reaches the value zero. If we take into account that the gravitational acceleration actually decreases as $1/r^2$ and therefore the scale height increases, the pressure decreases even more slowly at large distances and we can imagine that at very large distances from the star the gravitational force is no longer able to bind the material to the star against the gradient of the gas pressure which has to reach the very low interstellar value at large distances. At large distances we also have to take into account the decrease in temperature because the radiative flux $F_r$ also decreases as $1/r^2$. In radiative equilibrium the temperature $T$ is proportional to $F_r^{1/4}$ (see equation 6.31), so the temperature decreases at large distances roughly as $T \propto 1/\sqrt{r}$, which means the decrease in gravity outweighs the decrease in temperature, and in a rough approximation we can say that the scale height $H$ increases as $r^{3/2}$.

Chapman (1957) considered the case of a stellar corona in which the temperature stratification is determined by heat conduction and found

$$T = T_0 \left(\frac{r_0}{r}\right)^{2/7}. \qquad 16.5$$

The temperature then decreases more slowly than in the case of radiative equilibrium and the scale height increases somewhat faster. (In the presence of winds the temperature has to decrease faster because of the expansion of the wind.) If we insert Chapman's outer coronal temperature stratification into the hydrostatic equilibrium equation and remember that for a fully ionized hydrogen atmosphere with $n =$ number of protons

$$\rho = nm_{\mathrm{H}} \quad \text{and} \quad P_g = 2nkT, \tag{16.6}$$

we find

$$\frac{dP_g}{dr} = \frac{-GM}{r^2} nm_{\mathrm{H}}, \tag{16.7}$$

and

$$2k\frac{d}{dr}(nT) = 2kT_0 r_0^{2/7}\frac{d}{dr}(n/r^{2/7}) = \frac{-GM}{r^2}nm_{\mathrm{H}}, \tag{16.8}$$

or

$$\frac{d}{dr}\left(\frac{n}{r^{2/7}}\right) = -\frac{GMnm_{\mathrm{H}}}{2kT_0 r_0^{2/7}} r^{\frac{1}{2}}. \tag{16.9}$$

A solution of this equation is

$$n(r) = n(r_0) x^{2/7} e^{C[x^{-5/7} - 1]} \tag{16.10}$$

with

$$x = \frac{r}{r_0} \quad \text{and} \quad C = \frac{7}{5}\frac{GMm_{\mathrm{H}}}{2kT_0 r_0}. \tag{16.11}$$

For small distances one obtains with $x = 1 + (r - r_0)/r_0$ and $(r - r_0)/r_0 \ll 1$

$$n(r) = n(r_0)\exp\left(-\frac{GMm_{\mathrm{H}}}{2kT_0 r_0}\frac{r - r_0}{r_0}\right). \tag{16.12}$$

This is similar to equation (16.4), but the scale height $H$ is now replaced by

$$H' = \frac{2kT_0 r_0^2}{GMm_{\mathrm{H}}} = \frac{2R_g T_0 r_0^2}{GM}. \tag{16.13}$$

This new scale height is a constant, even though the dependence of $T$ and $g$ on $r$ have been taken into account. At the stellar surfaces, where $r = R$, we find that

$$\frac{H'}{H} = \frac{2R^2\mu}{R^2} = 2\mu \approx 1. \tag{16.14}$$

At the stellar surface the isothermal scale height $H$ describes the density stratification rather well.

For the temperature of the solar corona $T_0 = 2 \times 10^6$ K we calculate $H' \approx 60\,000$ km. Using the observed density of $n(r_0) \approx 10^6$ cm$^3$ for the outer corona, we find for the expected particle density at the distance of the Earth, namely, $d = 1.5 \times 10^{13}$ cm $= 1$ AU or $x = 2 \times 10^2$, $n(d) \approx 10^3$.

This is a surprisingly large density – much larger than is actually observed – telling us that the density in the far out layers of the solar corona decreases much faster than in hydrostatic equilibrium. The actual pressure gradient must therefore be much larger than in hydrostatic equilibrium and so must be too large to be balanced by the gravitational forces. This means that there is a net pressure force pushing the particles outward, away from the sun. The question then is why does such a large pressure gradient develop?

Let us now look at the gas pressure as obtained from the hydrostatic equilibrium density stratification equation (16.10), which comes out to be

$$P_g(r) = P_{g_0} \exp\left(\frac{7}{5}\frac{r_0}{H'}\,(x^{-5/7} - 1)\right). \qquad 16.15$$

Here we have also made use of equation (16.5).

For the pressure at very large distances we find

$$P_g(\infty) = P_{g_0} e^{-\frac{7}{5}r_0/H'}, \qquad 16.16$$

which means at infinity the pressure reaches a constant value. Because the temperature decreases with increasing distance, this means also that the density does not decrease to zero with increasing distance from the star in hydrostatic equilibrium. This in turn means that the total mass contained in this far out corona must be infinite.

All these derivations demonstrate that hydrostatic equilibrium is not possible at very large distances from the stars because the gravitational binding force is too small. For the solar coronal temperature of $2 \times 10^6$ K density equation (16.16) gives as the pressure at very large distances

$$P_{g_\odot}(\infty) \approx 10^{-10} \text{ dyn cm}^{-2},$$

while the actual interstellar gas pressure is of the order of $10^{-14}$ to $10^{-13}$ dyn cm$^{-2}$. In order to match this interstellar gas pressure the pressure gradient in the far out solar corona has to be larger than can be balanced by the solar gravitational force. The coronal gas is then accelerated outward by the excess pressure gradient, which leads to the solar wind. Once the gas at large distances starts streaming out, the gas at smaller distances has to follow. Otherwise, a vacuum would be created somewhere which would cause an even larger pressure gradient, forcing the gas to fill in this very low density region. The solar wind is therefore a necessary consequence of the impossibility to fulfill both the hydrostatic equilibrium equation and the boundary condition at infinity, where the stellar gas pressure has to become equal to the interstellar gas pressure.

Once we know that t' ɛ gas has to stream outwards, we can make use of the hydrodynamic equations in order to see which functions $\rho(r)$ and $v(r)$ are possible.

## 16.2.2 *Hydrodynamic theory of stellar winds*

We are looking for a steady state flow in the outer layers of the corona. In reality, the corona of the sun and probably also of other stars is rather inhomogeneous and the flow appears to be variable in time as well. But on average over the solar sphere and over long times the flow appears to be steady. We can therefore assume that the steady state solution gives us some insight and understanding of the time and space averaged flow conditions.

The hydrodynamic equation of motion for a spherically symmetric steady flow is (see, for instance, Mihalas, 1978, p. 516)

$$\rho v \frac{dv}{dr} = -\frac{dP_g}{dr} - \frac{GM}{r^2} \rho, \qquad\qquad 16.17$$

where $\rho v$ is the mass flow passing through $1\,\text{cm}^2$ per second, $v$ being the radial velocity component. If there is a constant mass flow going through all spherical shells, such that no mass is collected somewhere and no vacuum is created at another place, then we must require that

$$\rho v 4\pi r^2 = \text{const.}, \qquad\qquad 16.18$$

or

$$\frac{d}{dr}(\rho v r^2) = 0. \qquad\qquad 16.19$$

In order to relate $P_g$ and $\rho$ in equations (16.17) and (16.18), we need to know $T(r)$, which is determined by the energy equation, for which the solution is rather complicated to obtain. This difficulty can be circumvented by assuming a temperature stratification, as we did with equation (16.15). We can also assume a reasonable relation between $P_g$ and $\rho$. In the simplest case we can assume that $T = \text{const.} = T_0$. For the solar corona the temperature actually seems to vary rather little out to about three solar radii. With these simplifying assumptions equation (16.6) holds, if we consider a fully ionized atmosphere. The equation of motion now becomes

$$nm_{\text{H}} v \frac{dv}{dr} = -2kT \frac{dn}{dr} - nm_{\text{H}} \frac{GM}{r^2}. \qquad\qquad 16.20$$

Differentiation of (16.18) yields

$$\frac{d}{dr}(nvr^2) = 0 = vr^2 \frac{dn}{dr} + nr^2 \frac{dv}{dr} + 2nvr. \qquad\qquad 16.21$$

Solving for $dn/dr$ we obtain

$$-\frac{dn}{dr} = \frac{n}{v}\frac{dv}{dr} + \frac{2n}{r}. \qquad\qquad 16.22$$

This equation prescribes how the density (described by the number $n$ of particles per cm$^3$) must vary with $r$ for a given $v(r)$. We can rewrite equation (16.22) in the form

$$\frac{d \ln n}{dr} = -\frac{d \ln v}{dr} - \frac{2}{r}, \qquad\qquad 16.23$$

which shows even more clearly how the density variation depends on the velocity law. The density gradient and thereby also the pressure gradient is completely determined by the flow velocity.

Division of equation (16.20) by $n$ and replacing $d \ln n/dr$ by the expression in equation (16.23) yields the final equation of motion

$$m_H v^2 \frac{d \ln v}{dr} = 2kT\left(\frac{d \ln v}{dr} + \frac{2}{r}\right) - m_H \frac{GM}{r^2}, \qquad\qquad 16.24$$

which does not contain the density. The flow pattern is independent of the density. For a given $T$ and $M$ the flow pattern can be calculated uniquely. The density gradient is therefore also completely determined for a given $T$ and $M$.

(We must keep in mind, however, that we have not considered any other but gravitational and pressure forces. Magnetohydrodynamic forces have been neglected.)

Solving equation (16.24) for $d \ln v/dr$ we obtain

$$\frac{d \ln v}{dr}\left(v^2 - \frac{2kT}{m_H}\right) = \frac{4kT}{m_H r} - \frac{GM}{r^2}. \qquad\qquad 16.25$$

Since hydrostatic equilibrium is not possible, as we saw above, the star must find a solution of equation (16.25), which leads to positive velocities everywhere, including very large distances. On the other hand, if the star keeps existing, the velocities near the stellar surface must decrease to near zero. Otherwise, the star would dissolve in a short time. It turns out that there is just one solution of equation (16.25) which fulfils these conditions. Since the star is gravitationally bound, we must have close to the star

$$\frac{GM}{r^2} > \frac{4kT}{m_H r}. \qquad\qquad 16.26$$

The right-hand side of equation (16.25) is therefore negative close to the star. In order for the $d \ln v/dr$ to be positive the left-hand bracket must also be negative, which is the case if $v^2 < 2kT/m_H$. This means for small distances the velocities must be small, as we required.

For increasing distances from the star the term $GM/r^2$ decreases faster than the term $4kT/m_H r$. At some distance, $r = r_c$, with $r_c$ depending on $T$, the right-hand side will therefore become zero and then change sign. If at this point the left-hand bracket is still negative, then d ln $v/dr$ must also change sign, which means it must become negative. The velocity then starts decreasing at this point and finally becomes zero. This is not the physically required solution for which we are looking. For a finite velocity at large distances we must require that at $r = r_c$, also called the critical radius, the velocity must have increased enough such that at this point the left-hand bracket also becomes zero and then changes sign. Only if this is so do we find a positive velocity gradient everywhere and reach positive velocities at infinity. How does the star adjust so that this does actually happen? Suppose the velocity does not increase outwards. Then the material retains too high a density at this point and the pressure gradient between this point and the interstellar medium becomes very large. This large pressure gradient will push the material out and will increase the outward velocity. The material behind this point must follow. Otherwise, a large pressure gradient builds up behind the critical point, which then pushes this material out until the correct velocity distribution is achieved. In other words, the pressure gradient adjusts to the values which create the required velocity field.

Once we have calculated this singular solution $v(r)$ of equation (16.25), the density and pressure gradients can be determined from equations (16.23) and (16.20).

## 16.3    Theoretical properties of stellar winds

### 16.3.1    Critical velocities and temperatures

In Fig. 16.7 we show the dependence of the velocity $v(r)$ on the distance $r$, measured in units of $r_c$, which is the radius for which

$$v = v_c = \left(\frac{2kT}{m_H}\right)^{1/2}.$$

16.27

$r_c$ is called the critical radius. $v_c = c$ is the velocity of sound for isothermal sound waves. $r_c$ is therefore also called the sonic point. From Fig. 16.7 we find that at very large distances the final velocity, also called the terminal velocity, $v(\infty)$, reaches about three times the critical velocity $v_c$. For the solar coronal temperature $T = 2 \times 10^6$ K, we find $v_c = 182$ km s$^{-1}$ and $v(\infty) = 550$ km s$^{-1}$, which is just about the velocity which we observe in the solar wind, i.e., at a distance of one astronomical unit.

For $T = 30\,000$ K, as observed in luminous cool stars, we calculate $v_c = 22$ km s$^{-1}$ and $v(\infty) = 66$ km s$^{-1}$. Again, we find the correct order of magnitude. However, we must remember that we have assumed an isothermal corona. We must expect some cooling of the corona due to the expansion of the outflowing material.

For a realistic temperature determination we have to solve the energy equation. However, we do not know whether there is still mechanical energy deposited in the stellar winds beyond the sonic point, which has to be included in the energy equation.

### 16.3.2  Coronal temperatures

For the temperatures in the stellar coronae we find at the critical points (see the right-hand side of equation 16.25)

$$T_c = \frac{GM}{r_c^2} \frac{m_H r_c}{4k} = \frac{GM}{r_c} \frac{1}{4R_g}. \qquad 16.28$$

We clearly must have $r_c > R$, where $R$ is the stellar radius. Otherwise, the star would blow apart. We can therefore conclude that

$$T_c \lesssim \frac{GM}{R} \frac{1}{4R_g} \qquad 16.29$$

is an upper limit for coronae with Parker type stellar winds.

Expressing $M$ and $R$ in solar units, we find

$$T_c \lesssim \frac{M}{M_\odot} \frac{R_\odot}{R} \frac{GM_\odot}{R_\odot 4R_g} = \frac{M}{M_\odot} \frac{R_\odot}{R} 5.73 \times 10^6, \qquad 16.30$$

which tells us that for solar type coronae the temperature cannot be higher

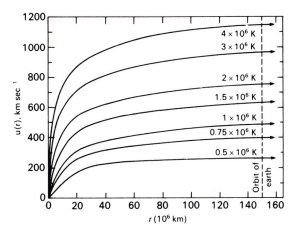

*Fig. 16.7.* The dependence of the wind velocity on the distance $r$ from the solar center is shown, where $r$ is measured in units of $10^6$ km. The velocities of these isothermal coronae depend on the coronal temperature, given on each curve. (From Hundhausen, 1972, p. 10.)

than about $6 \times 10^6$ K. Otherwise, the sonic point would move into the star and the mass loss would be tremendous. If the mass loss were that large the energy loss would be so large that there would not be enough mechanical energy available to make up for this loss and the corona would have to cool down, thereby reducing the stellar wind.

Observations of so-called RS CVn stars show, however, much higher coronal temperatures for these stars. We must conclude that in such stars the stellar winds are suppressed, possibly due to magnetic fields with large horizontal components.

### 16.3.3  Distances of critical points

Solving equation (16.28) for $r_c/R$ we can determine how far out we find the sonic point. The value of $r_c/R$ depends on $T_c$ and on $M/R$. We find

$$\frac{r_c}{R} = 5.7 \times 10^6 \frac{1}{T_c} \frac{M}{R} \qquad 16.31$$

where $M$ and $R$ are the mass and radius of the star in solar units.

For the solar values of $T_c = 2 \times 10^6$ K and $M = 1$, $R = 1$, we calculate

$$(r_c/R)_{\odot} = 2.8. \qquad 16.32$$

For this coronal temperature in the sun the sonic point should be at about three solar radii, but for a temperature of only $10^6$ K it would be at a distance of 5.5 $R_{\odot}$. The distance increases rather rapidly with decreasing temperature.

For very cool, luminous stars with $T_c \sim 10\,000$ K and with $M \approx 10$ $M_{\odot}$, $R \approx 100$ $R_{\odot}$, we derive

$$\frac{r_c}{R} = 57. \qquad 16.33$$

For these stars the sonic point is found to be very far away from the star. With $g = 30$ the pressure scale height is roughly $H = 5.5 \times 10^{10}$ cm, or about 1% of the stellar radius. At a distance of 50 stellar radii the density should have decreased to very low values indeed if this scale height were correct. Observations show, however, that these stars have much larger scale heights. The effective gravities must be much smaller than the purely gravitational one. The reason for this is still debated. For lower effective gravities the sonic point is moved closer to the stellar surface.

For hot, luminous stars with $M/R = 2$ and $T(\text{surface}) \approx 40\,000$ K we would find

$$r_c/R \approx 2.8, \qquad 16.34$$

which is similar to the sun. However, with a scale height of $6.6 \times 10^8$ cm, the distance corresponds to 2000 scale heights, which means the density at this point should be extremely small. In fact, the interstellar pressure could easily be matched for such a star. However, these stars are observed to have a strong mass loss. The density at the sonic point must therefore be much larger than calculated. This in turn means that the effective gravity must be much smaller than the gravitational one. There must be additional forces opposing the gravitational forces. For hot stars radiation pressure is quite important and is probably the origin of the observed strong winds, although it is still surprising that such large velocities can be achieved.

## 16.4    The Eddington limit

In our previous discussion of winds in hot stars we claimed that the forces exerted on the atoms and ions by photons can reduce the effective gravity in stars and can cause strong stellar winds. We may ask whether the radiative forces might ever be strong enough to destroy the whole star. This question has been studied already by Eddington.

When discussing diffusion in stars (see Volume 1, Chapter 15), we pointed out that each absorbed photon transfers a momentum of $hv/c$ in the direction of propagation to the absorbing particle. In an isotropic radiation field an equal number of photons come from all directions and the net force on all the absorbing atoms and ions is zero. In a stellar atmosphere the radiation field is not isotropic. We must always have a surplus of photons coming from the direction of the stellar interior such that there can be a net flux of energy from the inside out in order to replenish the energy lost at the surface. This energy flux is described by

$$\pi F = \pi \int_0^\infty F_v \, dv.$$

Only the forces due to this surplus of photons coming from the direction of the center and flowing radially outwards are important because, as we saw, the forces due to the isotropic part of the radiation field cancel.

In Chapter 9 we found that the transfer of momentum to the particles due to the absorption of photons can be described by an outward directed radiative acceleration $g_r$, which is given by

$$g_r = \frac{1}{c} \int_0^\infty (\kappa_{v,gr} + \sigma_{v,gr}) \pi F_v \, dv. \qquad 16.35$$

The net acceleration $g_{\text{eff}}$ is given by

$$g_{\text{eff}} = g_g - g_r,$$    16.36

where $g_g$ is the gravitational acceleration.

The star can only hold together as long as $g_{\text{eff}}$ is positive, which means as long as there is still a force directed towards the center of the star. For stability of the star we must therefore require

$$g_r < g_g.$$    16.37

For very low densities, as expected in the outer layers for stars with a low $g_{\text{eff}}$, the absorption and scattering coefficients are dominated by the Thomson scattering by free electrons. We can therefore obtain a rough estimate for the radiative acceleration by assuming that $\kappa_{v,gr} + \sigma_{v,gr} \approx \sigma_{v,el,gr}$. There may be, of course, lines superimposed on the Thomson scattering, or for high temperatures, UV continuum absorption edges may be larger than $\sigma_v$. The true radiative acceleration may therefore be larger than obtained by considering electron scattering only, but using $\sigma_{v,el}$ still gives us a first estimate. Because $\sigma_{v,el}$ is independent of frequency, we can take it out of the integral in equation (16.35) and derive

$$g_r = \frac{\sigma_{v,gr}}{c} \int_0^{\infty} \pi F_v \, dv = \frac{\sigma_{v,gr}}{c} \pi F = \frac{\sigma_{v,gr}}{c} \sigma T_{\text{eff}}^4.$$    16.38

For stability of the star we must then require

$$g_g = \frac{GM}{R^2} > g_r \approx \frac{\sigma_{v,gr}}{c} \sigma T_{\text{eff}}^4,$$    16.39

or

$$GM > \frac{\sigma_{v,gr}}{c} \sigma T_{\text{eff}}^4 R^2 = \frac{\sigma_{v,gr}}{c} \frac{L}{4\pi},$$    16.40

because $L = \sigma T_{\text{eff}}^4 4\pi R^2$.

This can be rewritten as

$$L < \frac{GM 4\pi c}{\sigma_{v,gr}} = \frac{M}{M_{\odot}} \frac{GM_{\odot} 4\pi c}{\sigma_{v,gr}} = \frac{M}{M_{\odot}} \times 2 \times 10^{38} = \frac{M}{M_{\odot}} 5 \times 10^4 L_{\odot},$$    16.41

where we have used $\sigma_{v,gr} = 0.25$.

For a star with one solar mass, the maximum possible luminosity turns out to be $L(\text{max}) = 5 \times 10^4 L_{\odot}$, or $M_{\text{bol}} \geqslant -7.0$. For a star with 40 solar masses the limit is $M_{\text{bol}} \geqslant -11.0$. This is brighter than the magnitude for the brightest stars seen in our galaxy. If the luminosity of a star approaches the limit, then its $g_{\text{eff}}$ is already much smaller than its $g_g$. If line absorption is also important, then the maximum luminosity is reduced correspondingly, which means it can be reduced by up to a factor of 5 to 10.

# Problems

**Chapter 1**

1. With a photometer you measure the energy fluxes of a star whose unknown colors you want to determine. You also measure the fluxes of the standard star Vega ($\alpha$ Lyrae). For simplicity you measure both stars at the same zenith distance. For each filter the photometer readings are proportional to the apparent fluxes. For the different filters you have the following readings per unit of time (after background subtraction):

| | U | B | V |
|---|---|---|---|
| Unknown star | 2000 | 3000 | 2500 |
| $\alpha$ Lyrae | 5000 | 7500 | 6250 |

What are the apparent magnitudes of the unknown star?
What are the $U - B$ and $B - V$ colors of the star?
Assuming it is a main sequence star, what is its spectral type?
What are the absolute UBV magnitudes of the star?

2. You want to determine the unknown apparent blue magnitude of a star. You have determined the photometer readings for Vega which are proportional to the fluxes received when Vega was at a zenith distance of $20°$. You can measure the apparent fluxes for your program star only at zenith distances of $\delta = 30°$ and $\delta = 45°$.

Photometer readings per unit of time (after background subtraction):

| $\delta$ | Star | Vega |
|---|---|---|
| $20°$ | | 2000 |
| $30°$ | 800 | ? |
| $45°$ | 710 | ? |

Determine the correction factors for the extinction in the Earth's atmosphere. What is the apparent blue magnitude of the star?

3. The absolute visual magnitude of the sun if $M_V = 4.82$. The sun is at a distance of 1 astronomical unit $= 1.49 \times 10^{13}$ cm. What is the apparent visual magnitude of the sun?

## Chapter 3

1. The amount of energy we receive from the sun per $cm^2$ s above the Earth's atmosphere is called the solar constant $S$. It is $S = \pi f_\odot = 1.38 \times 10^6$ erg cm$^{-2}$ s$^{-1}$. Measure the angular radius of the sun and calculate $\pi F$, the surface flux on the sun. Calculate the effective temperature of the sun.

2. (a) The apparent visual magnitude of the K0 III star $\alpha$ Boo (Arcturus) is $m_v = -0.04$ and $B - V = 1.23$. The bolometric correction is about $BC = 0.30$. What is the apparent bolometric magnitude of the star?
   (b) The apparent bolometric magnitude of the sun is $m_{bol\odot} = -26.82$. The angular diameter of $\alpha$ Boo was measured with the Michelson interferometer to be $\alpha = 0.02$ arcsec. Calculate the effective temperature of $\alpha$ Boo.
   (c) The trigonometric parallax of $\alpha$ Boo was measured to be $\pi = 0.09$ arcsec. What is the radius of $\alpha$ Boo?

3. Show that the Planck function

$$B_\lambda = \frac{2hc^2}{\lambda^5} \frac{1}{e^{hc/\lambda kT} - 1}$$

has its maximum at a wavelength $\lambda_{max}(T)$ for which $\lambda_{max} T = $ const.
   (a) For a star with an apparent visual magnitude $m_V = 0$ we receive in the visual wavelength band an energy of $\pi f_v = 3.6 \times 10^{-9}$ erg cm$^{-2}$ s$^{-1}$ Å$^{-1}$ above the Earth's atmosphere. How much energy do we receive from Sirius in the visual? Sirius has $m_v = -1.45/$
   (b) Sirius is an A1 V star for which the bolometric correction is $BC = 0.30$. What is the apparent bolometric magnitude of Sirius?
   (c) The apparent bolometric magnitude of the sun is $m_{bol\odot} = -26.82$. What is the total amount of energy $\pi f$ which we receive from Sirius above the Earth's atmosphere per $cm^2$ s?
   (d) The angular diameter of Sirius was measured to be $\alpha = 6.12 \times 10^{-3}$ arcsec. What is the $T_{eff}$ of Sirius? The Stefan–Boltzmann constant is $\sigma = 5.67 \times 10^{-5}$ erg deg$^{-4}$.

## Chapter 4

1. You have a well-insulated box filled with sodium vapor. The box has been left for a long time. Suppose you punch a small hole into the box and observe the energy distribution around the strong yellow sodium lines at $\lambda = 5896$ Å and 5890 Å. What would the energy distribution look like? Would you see spectral lines? If so, would you see absorption or emission lines?

2. Many early B stars have gaseous shells around them, which are optically thin. Suppose this shell expands with a velocity of about 50 km s$^{-1}$. How would the line profile of a strong Si III line look in the stellar spectrum?

## Chapter 5

1. Does the foreshortening factor $\cos \delta$ discussed in equation (4.2) for light leaving $1 \, \text{cm}^2$ of the stellar surface under an angle $\delta$ lead to a limb darkening which we have omitted in our discussion?

2. We have been looking only at what happens in the light beam under consideration. There are many photons scattered from all other directions into this beam. Did we consider those too or were they omitted? If we did consider them, how did we do this?

3. The source function $S(\tau_\lambda)$ is approximated by

$$S_\lambda(\tau_\lambda) = \sum_{i=0}^{i=n} a_{\lambda i} \tau_\lambda^i.$$

Show that $I_\lambda(0, \vartheta)$ is then given by

$$I_\lambda(0, \vartheta) = \sum A_{\lambda i} \cos^i \vartheta$$

where $A_{\lambda i} = \alpha_{\lambda i} \cdot i!$. Remember that

$$\int_0^\infty x^i e^{-x} \, dx = i!.$$

4. We approximate the source function by

$$S_\lambda(\tau_\lambda) = a_\lambda + b_\lambda \tau_\lambda + c_\lambda \tau_\lambda^2.$$

Calculate $F_\lambda(0)$.

5. The center to limb variation for the solar radiation at 5010 Å has been measured to be

$$\frac{I_\lambda(0, \vartheta)}{I_\lambda(0, 0)} = 0.2593 + 0.8724 \cos \vartheta - 0.1336 \cos^2 \vartheta.$$

$I_\lambda(0, 0)$ for 5010 Å $= 4.05 \times 10^{14} \, \text{erg cm}^{-2} \, \text{s}^{-1}$ per $\Delta \lambda = 1 \, \text{cm}$.
Calculate $S_\lambda(\tau_\lambda)$.
Assume $S_\lambda = B_\lambda(T(\tau_\lambda))$. Calculate $T(\tau_\lambda)$, where $\tau_\lambda$ is the optical depth at 5010 Å. Plot $S_\lambda(\tau_\lambda)$ and $T(\tau_\lambda)$.

## Chapter 6

1. The measured center to limit variation of the solar intensity is for the different wavelengths
$$I_\lambda(0, \vartheta)/I_\lambda(0, 0) = a_0 + a_1 \cos \vartheta + 2a_2 \cos^2 \vartheta.$$

| $\lambda$ [Å] | $a_0$ | $a_1$ | $2a_2$ | $I_\lambda(0, 0)^*$ |
|---|---|---|---|---|
| 3737 | 0.1435 | 0.9481 | −0.0920 | $42.0 \times 10^{13}$ |
| 4260 | 0.1754 | 0.9740 | −0.1525 | $44.9 \times 10^{13}$ |
| 5010 | 0.2593 | 0.8724 | −0.1336 | $40.3 \times 10^{13}$ |
| 6990 | 0.4128 | 0.7525 | −0.1761 | $25.0 \times 10^{13}$ |
| 8660 | 0.5141 | 0.6497 | −0.1657 | $15.5 \times 10^{13}$ |
| 12 250 | 0.5969 | 0.5667 | −0.1646 | $7.7 \times 10^{13}$ |
| 16 550 | 0.6894 | 0.4563 | −0.1472 | $3.6 \times 10^{13}$ |
| 20 970 | 0.7249 | 0.4100 | −0.1360 | $1.6 \times 10^{13}$ |

* Units: erg cm$^{-2}$ s$^{-1}$ cm$^{-1}$

Calculate $T(\tau_\lambda)$ for the different wavelengths $\lambda$.
Plot $T(\tau_\lambda)$ for the different wavelengths $\lambda$.
Calculate $\log \tau_\lambda(\lambda)$ for $T = 6000$ and $T = 6500$ K.

2. (a) Verify the surface condition for a grey atmosphere, namely, $J(0) = \frac{1}{2}F(0)$, assuming isotropic radiation for the outer half sphere and no incoming radiation.

   (b) Verify the Eddington approximation for this case of isotropy in a half sphere and no radiation in the other half sphere.

## Chapter 7

1. For a pure hydrogen gas with a gas pressure of $P_g = 10^3$ dyn cm$^{-2}$ and a temperature $T = 10\,080$ K, calculate the ratio H$^+$/H and the electron pressure $P_e = n_e kT$. Remember that $n_e = $ H$^+$ and $P_g = nkT$, with $n = e + $ H$^+$ + H. $\chi_{ion} = 13.6$ eV.

2. For a pure helium gas with a gas pressure $P_g = 10^3$ dyn cm$^{-2}$ and $T = 15\,000$ K, calculate He$^{2+}$/He$^+$, He$^+$/He, and $P_e$. Remember $n = n_e + $ He$^{2+}$ + He$^+$ + He. $\chi_{ion}$(He) $= 24.58$ eV; $\chi_{ion}$(He$^+$) $= 54.4$ eV. Here $\chi_{ion}$(He$^+$) is the energy needed in order to remove the additional electron from the He$^+$ ion to make He$^{2+}$.

3. The absorption coefficient per hydrogen atom in the level with main quantum number $n$ is given by
$$a_n = \frac{1}{n^2} \times \frac{64}{3\sqrt{3}} \times \frac{\pi^4}{v^3} \times \frac{Z^4 m e^{10}}{ch^6 n^3},$$
where

   $v = c/\lambda$, $c = $ velocity of light, $c = 3 \times 10^{10}$ cm s$^{-1}$
   $m = $ electron mass $= 9.105 \times 10^{-28}$ g
   $h = $ Planck's constant $= 6.624 \times 10^{-27}$ erg s
   $e = $ charge of electron $= 4.8024 \times 10^{-10}$ electrostatic units
   $Z = $ nuclear charge $+ 1 - $ number of electrons in atom or ion
   $Z = 1$ for H and He, $Z = 2$ for He$^+$.

   Calculate the hydrogen bound free absorption coefficient per atom (i.e., $N_H(n=1) = 1$) for the levels $n \leqslant 4$. Assume $T = 5800$ K, corresponding to the solar photosphere.

4. For the sun compare the number of hydrogen atoms absorbing at $\lambda = 3300$ Å with the number of H$^-$ ions absorbing at that wavelength. Assume a temperature $T = 5800$ K and an electron pressure $\log P_e = 1.3$.

5. For the sun calculate the number of Si I atoms absorbing at 1600 Å and compare with the number of hydrogen atoms absorbing at this wavelength. $\chi_{ion}$ (Si) $= 8.15$ V. For this estimate use $U^+ \approx u$ and $g_x \approx g_0$.

## Chapter 8

1. Just for the fun of it assume that only the hydrogen absorption is important for the sun, i.e., neglect H$^-$, metals, electron scattering and Rayleigh scattering. Calculate the emergent energy distribution in the continuum. For simplicity, use

a grey atmosphere temperature distribution and a depth independent absorption coefficient, i.e., use $\kappa(H)$ for $T_{eff} = 5800$ K. Plot $F_\lambda(0)$ for the surface.

## Chapter 9

1. Assume that the temperature and pressure dependence of the continuous absorption coefficient $\bar\kappa_{gr}$ can be approximately described by $\bar\kappa_{gr} = \kappa_0 \tau$. This is a reasonably good approximation for temperatures between 6000 K and 10 000 K.

   Calculate how the gas pressure increases with $\bar\tau$ in a given atmosphere.

2. Calculate the gravitational acceleration for the sun. $M_\odot = 2 \times 10^{33}$ g, $R_\odot = 6.96 \times 10^{10}$ cm, $G = 6.68 \times 10^{-8}$ [cgs]. Calculate the gas pressure at the optical depth $\bar\tau = \frac{2}{3}$. Assume $\bar\kappa_{gr} = \bar\kappa_{gr_0} P_g$ with $\log \bar\kappa_{gr_0} = -5.4$.

## Chapter 10

1. Explain why spectral lines in the red are generally weaker than the lines in the blue and ultraviolet wavelengths.

2. For a grey atmosphere calculate the maximum line depth for a spectral line at 3000 Å, at 6000 Å, and at 10 000 Å. Use $T_{eff} = 5800$ K.

3. Derive how the line depth of an optically thin line of neutral iron, i.e., the Fe I line depth, will change with decreasing metal abundances in solar type stars. Derive how the line depth for an optically thin $Fe^+$ line will change with decreasing abundances of the heavy elements. Assume that the abundances of all heavy elements are reduced by the same factor.

4. An early F star has a temperature of 7800 K. In the atmosphere we find a microturbulence with a reference velocity of $\xi_t = 4$ km s$^{-1}$. What is the half width of an optically thin Fe line in the star? How large is the thermal half width?

5. A spectral line corresponds to a transition between energy levels with quantum numbers $n$ and $m$, which have natural widths corresponding to damping constants $\gamma_n$ and $\gamma_m$. Which damping constant do you have to use for the calculation of the damping profile of this line?

6. A star is rotating with an equatorial velocity of 50 km s$^{-1}$. This rotation broadens its lines. Estimate the half widths of the lines due to this rotation. Compare this half width with the thermal Doppler half widths of an $Fe^+$ line in an A0 V star.

7. If in a star the half width of the line absorption coefficient of an Fe II line is increased by a factor of 2 due to microturbulence, by how much is its equivalent width increased when it is on the flat part of the curve of growth?

## Chapter 11

1. In Fig. 1 we show the spectra of the star $\alpha$ Lyr, which has a spectral type A0 V, and of the star $\eta$ Leo, which has a spectral type A0 I. Determine the electron densities for these two stars, making use of the Inglis–Teller formula.

2. How does the depth in the hydrogen line wings change with decreasing metal abundances
   (a) for solar type stars?
   (b) for A stars?
   Why are the hydrogen lines such a good means for the temperature determination?

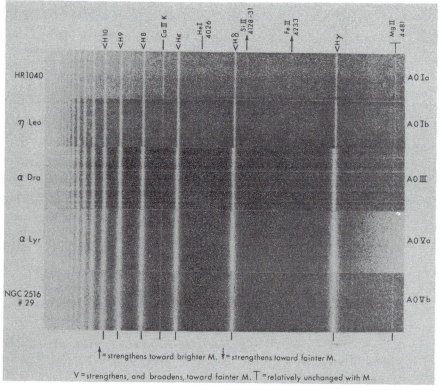

*Fig. 1.* Spectra of A stars with different luminosity classes are shown. For higher luminosity stars the hydrogen lines are narrower. More Balmer lines can be seen separately around 3800 Å for the higher luminosity stars. (See Chapter 11, problem 1. From Morgan, Abt and Tapscott, 1978.)

3. How does the depth in the hydrogen line wings change for increasing He abundance. That is for decreasing hydrogen abundance,
   (a) in solar type stars?
   (b) in A stars?

4. How does the Balmer discontinuity change with decreasing abundances of heavy elements
   (a) for F stars?
   (b) for B stars?

5. How does the Balmer discontinuity change with increasing He abundance
   (a) for F stars?
   (b) for B stars?

## Chapter 12

1. In Table 1 we give equivalent widths $W_\lambda$ of Fe I and Fe II lines and of Ti I and Ti II lines as measured for the sun. In the same table we also give the $gf$ values for these lines as measured in the laboratory.
   (a) Determine the excitation temperature for the sun for the energy levels involved.
   (b) Determine the Doppler width $\Delta\lambda_D$ for the Fe and Ti lines, and determine the microturbulent velocity $\xi_t$.

Table 1. *Equivalent widths and oscillator strengths for some solar spectral lines*

| Line $\lambda$ (Å) | $W_\lambda$ (mÅ) | log $gf$ | Line $\lambda$ (Å) | $W_\lambda$ (mÅ) | log $gf$ | Line $\lambda$ (Å) | $W_\lambda$ (mÅ) | log $gf$ |
|---|---|---|---|---|---|---|---|---|
| **Fe I** | | | **Ti I** | | | **Fe II** | | |
| Multiplet 1, $\chi_{s,l}=0.05$ eV | | | Multiplet 3, $\chi_{s,l}=0.03$ eV | | | Multiplet 37, $\chi_{s,l}=2.83$ eV | | |
| 5166.29 | 115 | −3.68 | 5460.51 | 8.5 | −2.36 | 4472.93 | 39 | −4.76 |
| 5247.06 | 59 | −4.50 | 5426.26 | 5.5 | −2.48 | 4489.19 | 61 | −3.69 |
| 5254.96 | 92 | −4.23 | 5490.84 | 2.5 | −2.67 | 4491.41 | 66 | −2.78 |
| 5110.41 | 126 | −3.34 | | | | 4555.89 | 77 | −2.42 |
| 5168.90 | 114 | −3.49 | Multiplet 4, $\chi_{s,l}=0.03$ eV | | | 4582.84 | 49 | −3.17 |
| 5225.53 | 68 | −4.26 | 5210.39 | 86 | −0.90 | 4520.23 | 69 | −3.20 |
| | | | 5192.97 | 80 | −0.96 | 4534.17 | 53 | −3.33 |
| Multiplet 15, $\chi_{s,l}=0.97$ eV | | | 5173.74 | 67 | −1.06 | | | |
| 5328.05 | 375 | −1.43 | 5219.71 | 25 | −1.90 | Multiplet 38, $\chi_{s,l}=2.82$ eV | | |
| 5405.78 | 266 | −1.78 | 5152.20 | 38 | −1.73 | 4508.28 | 74 | −2.42 |
| 5397.15 | 239 | −1.85 | 5147.48 | 36 | −1.71 | 4576.34 | 56 | −2.89 |
| 5429.70 | 285 | −1.76 | | | | 4620.51 | 47 | −3.13 |
| 5446.92 | 238 | −1.86 | Multiplet 38, $\chi_{s,l}=0.82$ eV | | | 4541.52 | 58 | −2.93 |
| 5455.61 | 219 | −2.01 | 4981.73 | 112 | +0.57 | | | |
| | | | 4991.07 | 102 | +0.45 | **Ti II** | | |
| Multiplet 66, $\chi_{s,l}=2.20$ eV | | | 4999.50 | 104 | +0.38 | Multiplets 18, 19, 20, | | |
| 5145.10 | 44 | −2.40 | 5016.16 | 60 | −0.44 | $\chi_{s,l}=1.08$ eV | | |
| 5131.48 | 72 | −1.92 | 5020.03 | 86 | −0.29 | 4469.16 | 49 | −3.00 |
| 5098.70 | 102 | −1.40 | 5022.87 | 72 | −0.30 | 4493.53 | 26 | −3.62 |
| 5079.23 | 100 | −1.45 | 5024.84 | 62 | −0.47 | 4395.03 | 135 | −0.55 |
| 5250.65 | 104 | −1.52 | 5045.40 | 10 | −1.49 | 4443.80 | 124 | −0.73 |
| 5198.71 | 87 | −1.50 | 5043.58 | 14 | −1.30 | 4450.49 | 79 | −1.56 |
| | | | 5040.64 | 16 | −1.37 | | | |
| Multiplet 687, $\chi_{s,l}=3.40$ eV | | | | | | Multiplets 50, 51, 60, | | |
| 4966.10 | 114 | −0.30 | Multiplet 183, $\chi_{s,l}=2.09$ eV | | | $\chi_{s,l}=1.22$ eV | | |
| 4946.39 | 113 | −0.74 | 5224.30 | 36 | +0.42 | 4533.97 | 109 | −0.70 |
| 4882.15 | 70 | −1.10 | 5224.56 | 68 | −0.03 | 4563.76 | 120 | −0.86 |
| 4963.65 | 48 | −1.21 | 5223.62 | 11 | −0.09 | 4568.30 | 25 | −3.45 |
| 4875.90 | 55 | −1.39 | 5222.69 | 23 | −0.05 | 4589.96 | 70 | −1.72 |
| 4855.68 | 60 | −1.33 | 5263.48 | 13 | −0.27 | 4432.09 | 15 | −3.55 |
| 4843.16 | 67 | −1.30 | 5247.29 | 10 | −0.15 | 4399.77 | 115 | −1.32 |
| 4838.52 | 51 | −1.39 | 5186.33 | 7 | −0.36 | 4394.06 | 72 | −1.71 |
| 5039.26 | 73 | −0.89 | 5194.04 | 10 | −0.08 | 4418.34 | 70 | −2.24 |
| 5002.80 | 85 | −1.03 | 5201.10 | 11 | −0.22 | | | |
| 4950.11 | 76 | −1.08 | 5207.85 | 8 | −0.16 | Multiplet 105, $\chi_{s,l}=2.59$ eV | | |
| 4907.74 | 61 | −1.33 | | | | 4386.85 | 59 | −0.79 |
| | | | | | | | | |
| | | | | | | Multiplet 115, $\chi_{s,l}=3.10$ eV | | |
| | | | | | | 4488.32 | 45 | −0.62 |

(c) From the Doppler widths determined for the Fe and the Ti lines try to determine the kinetic temperature for the sun. (You will see that there are problems doing it.)

(d) Determine the relative abundances of Ti I and Fe I in the sun.

(e) Determine the relative abundances of Fe II and Ti II in the sun.

(f) Try to determine the ionization temperature and the electron pressure from the ionization equilibria for these two elements. (You will see that there are problems.) $\chi_{ion}$ (Fe) = 7.90 eV, $\chi_{ion}$ (Ti) = 6.83 eV, $u$(Fe) = 31, $u^+$(Fe) = 47, $u$(Ti) = 34, $u^+$ (Ti) = 56.

(g) Assume $T_{ionization} = T_{excitation}$ and determine the electron pressure in the solar atmosphere from the ionization equilibria of Ti and Fe.

(h) Estimate the gas pressure in the solar photosphere, knowing the relative abundances of heavy elements and hydrogen.

## Chapter 13

1. The 21 cm line of the hydrogen atom is a transition between two energy states for which the magnetic moment orientation of the proton with respect to the magnetic moment of the electron is different. The transition probability is extremely small, namely, $A_{u,l} = 2.84 \times 10^{-15}$ s$^{-1}$. The cross-section for collisional de-excitation is $C_{u,l} \sim 7 \times 10^{-15}$ cm$^2$.

Determine the excitation temperature of the upper level in the interstellar medium with a kinetic temperature of 100 K and a density of about 1 atom per cm$^3$.

## Chapter 14

1. Calculate the radiative gradient for the solar convection zone in a layer where $T = 10\,000$ K and log $P_g = 5.33$.

2. For any given layer in the hydrogen convection zone give the largest possible ratio of $F_{conv}/F_{rad}$, or in other words, the smallest value for $F_{rad}/\sigma T^4_{eff}$.

3. In the deep interior of stars the absorption coefficient $\bar{\kappa}_{gr}$ decreases with increasing temperature approximately as $\bar{\kappa}_{gr} \propto T^{-3.5}$ (Kramers' opacity law). In layers where there is no nuclear energy generation can there be convective instability?

4. Calculate the convective energy transport in a layer in the solar hydrogen convection zone where $T = 10\,000$ K and log $P_g = 5.33$, log $\bar{\kappa}_{gr} = 1.65$, $\nabla_{ad} = 0.15$, $c_{p/at} = 9k$, $\mu \approx 1.44$.

Calculate the convective energy transport in a layer in an early F star convection zone where $T = 10\,000$ K and log $P_g = 4.46$, log $\bar{\kappa}_{gr} = 1.45$, $\nabla_{ad} = 0.11$, $c_{p/at} \approx 20k$, $\mu = 1.30$.

## Chapter 15

1. The height of the solar chromosphere is about 2000 km. After the second contact during total solar eclipses this layer is slowly covered by the shadow of the moon. Calculate how long some of the chromospheric spectrum can be seen. (We cannot see the spectrum before second contact because the sunlight from the disc is too bright.)

2. Calculate the temperature stratification in the lower part of a transition region between chromosphere and corona, assuming a mechanical energy flux of $10^5$ erg cm$^{-2}$ s$^{-1}$ and a damping length $\lambda = 1.5 \times 10^6 T^{0.5}$ cm. Use $g_{\text{eff}} = g$ with $g = 2.7 \times 10^4$ cm s$^{-2}$.

3. Calculate the temperature stratification in the upper part of a transition layer, assuming a constant conductive flux $F_c = 10^5$ erg cm$^{-2}$ s$^{-1}$.

4. In red giant chromospheres the electron densities are about $n_e = 10^8$ cm$^{-3}$. If the temperature is 10 000 K and $g_{\text{eff}} = g$, how large would be the density at the sonic point for such a star? How large would the mass loss be if the velocity were 100 km s$^{-1}$? Assume that the red giant has a mass $M = 1.5$ M$_\odot$ and a radius of 200 R$_\odot$.

5. Calculate the momentum transferred to a C IV ion by the absorption of one photon with $\lambda = 1550$ Å (C IV resonance lines). How large will the velocity change of the ion be? The abundance of C/H in stellar atmospheres is about $10^{-3}$. If the momentum of the C IV ions is distributed over all ions in the wind, how large would be the wind velocity if all C IV absorb one 1550 Å photon? Assume that all carbon ions are C IV ions.

6. (a) For the calculation of the Eddington luminosity limit for stars we have assumed that $\sigma_v + \kappa_v \approx \sigma_v = $ const. In reality there are a large number of absorption lines. Assume that 20% of the spectrum is covered by strong absorption lines.

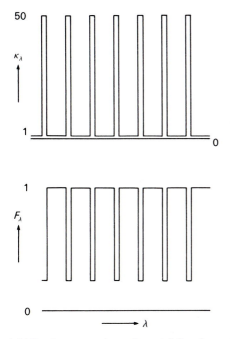

*Fig. 2.* (a) The frequency dependence of the absorption coefficient for our picket fence model in Chapter 15, problem 6. (b) The assumed frequency dependence of the flux in our picket fence model of problem 6.

For simplicity assume that these lines are evenly distributed over the spectrum and that within these lines $\kappa_L/\sigma_v = $ const. $= 50$. See Fig. 2. In such strong absorption lines the flux is reduced, assume that the flux in the lines is 25% of the continuum flux (see Fig. 2).

(i) By how much will the gravitational acceleration $g_r$ be increased due to these lines?

(ii) What would be the luminosity limit for such stars for $M = 1\,M_\odot$ and $M = 50\,M_\odot$?

(b) In the Large Magellanic Cloud (LMC) stars the abundances of the heavy elements are reduced by about a factor of 5 (the exact number is still uncertain) as compared to galactic stars. Suppose the line absorption coefficients in the above model (picket fence model) are reduced by this factor.

(i) By which factor is the radiative acceleration $g_r$ changed as compared to galactic stars?

(ii) What would be the luminosity limit for a $50\,M_\odot$ star in the LMC?

# Appendix

*LTE Model Stellar Atmospheres*[a]

| | | $T_{\text{eff}} = 5500$ K | | |
| --- | --- | --- | --- | --- |
| $\log \bar{\tau}$ | $T$ (K) | $\log P_g$ (dyn cm$^{-2}$) | $\log \bar{\kappa}_R$ (cm$^2$ g$^{-1}$) | $\log n_e$ (cm$^{-3}$) |
| $-3.0$ | 4251 | 3.50 | $-1.95$ | 11.56 |
| $-2.5$ | 4380 | 3.86 | $-1.64$ | 11.85 |
| $-2.0$ | 4491 | 4.14 | $-1.40$ | 12.15 |
| $-1.5$ | 4630 | 4.41 | $-1.15$ | 12.42 |
| $-1.0$ | 4842 | 4.70 | $-0.86$ | 12.75 |
| $-0.9$ | 4894 | 4.73 | $-0.85$ | 12.77 |
| $-0.8$ | 4961 | 4.79 | $-0.80$ | 12.82 |
| $-0.7$ | 5040 | 4.84 | $-0.75$ | 12.89 |
| $-0.6$ | 5126 | 4.90 | $-0.71$ | 12.96 |
| $-0.5$ | 5224 | 4.95 | $-0.65$ | 13.03 |
| $-0.4$ | 5335 | 4.99 | $-0.60$ | 13.10 |
| $-0.3$ | 5465 | 5.04 | $-0.55$ | 13.18 |
| $-0.2$ | 5612 | 5.09 | $-0.47$ | 13.28 |
| $-0.1$ | 5770 | 5.14 | $-0.38$ | 13.40 |
| 0 | 5975 | 5.18 | $-0.26$ | 13.54 |
| 0.1 | 6218 | 5.22 | $-0.11$ | 13.73 |
| 0.2 | 6465 | 5.25 | $+0.04$ | 13.93 |
| 0.3 | 6675 | 5.27 | $+0.18$ | 14.10 |
| 0.5 | 7098 | 5.31 | $+0.44$ | 14.41 |
| 0.7 | 7453 | 5.34 | $+0.65$ | 14.64 |
| 1.0 | 7913 | 5.39 | $+0.92$ | 14.95 |

$T_{\text{eff}} = 5500$ K
$\log g = 1.5,\ \log Z/Z_\odot = 0$

| $\log \bar{\tau}$ | $T$ (K) | $\log P_g$ (dyn cm$^{-2}$) | $\log \bar{\kappa}_R$ (cm$^2$ g$^{-1}$) | $\log n_e$ (cm$^{-3}$) |
|---|---|---|---|---|
| −3.0 | 4239 | 1.89 | −3.25 | 10.13 |
| −2.5 | 4384 | 2.21 | −2.99 | 10.44 |
| −2.0 | 4507 | 2.50 | −2.76 | 10.71 |
| −1.5 | 4651 | 2.78 | −2.51 | 11.02 |
| −1.0 | 4857 | 3.04 | −2.25 | 11.32 |
| −0.9 | 4910 | 3.09 | −2.19 | 11.40 |
| −0.8 | 4978 | 3.14 | −2.12 | 11.48 |
| −0.7 | 5057 | 3.19 | −2.05 | 11.56 |
| −0.6 | 5145 | 3.24 | −1.97 | 11.65 |
| −0.5 | 5245 | 3.28 | −1.88 | 11.76 |
| −0.4 | 5350 | 3.32 | −1.77 | 11.89 |
| −0.3 | 5467 | 3.35 | −1.66 | 12.02 |
| −0.2 | 5610 | 3.38 | −1.53 | 12.18 |
| −0.1 | 5780 | 3.41 | −1.38 | 12.35 |
| 0 | 5950 | 3.43 | −1.22 | 12.53 |
| 0.1 | 6135 | 3.45 | −1.05 | 12.71 |
| 0.2 | 6376 | 3.46 | −0.87 | 12.92 |
| 0.3 | 6665 | 3.47 | −0.64 | 13.17 |
| 0.5 | 7320 | 3.48 | −0.13 | 13.63 |
| 0.7 | 7700 | 3.49 | +0.15 | 13.86 |
| 1.0 | 8029 | 3.50 | +0.45 | 14.09 |

$T_{\text{eff}} = 5500$
$\log g = 4.5,\ \log Z/Z_\odot = -2$

| $\log \bar{\tau}$ | $T$ (K) | $\log P_g$ (dyn cm$^{-2}$) | $\log \bar{\kappa}_R$ (cm$^2$ g$^{-1}$) | $\log n_e$ (cm$^{-3}$) |
|---|---|---|---|---|
| −3.0 | 4447 | 4.29 | −2.54 | 11.02 |
| −2.5 | 4487 | 4.57 | −2.33 | 11.24 |
| −2.0 | 4536 | 4.85 | −2.11 | 11.48 |
| −1.5 | 4626 | 5.13 | −1.85 | 11.76 |
| −1.0 | 4827 | 5.37 | −1.50 | 12.15 |
| −0.9 | 4897 | 5.41 | −1.40 | 12.27 |
| −0.8 | 4980 | 5.45 | −1.29 | 12.40 |
| −0.7 | 5078 | 5.48 | −1.17 | 12.54 |
| −0.6 | 5200 | 5.51 | −1.04 | 12.69 |
| −0.5 | 5330 | 5.53 | −0.90 | 12.85 |
| −0.4 | 5458 | 5.55 | −0.77 | 13.03 |
| −0.3 | 5589 | 5.57 | −0.64 | 13.19 |
| −0.2 | 5717 | 5.59 | −0.52 | 13.34 |
| −0.1 | 5830 | 5.61 | −0.40 | 13.47 |
| 0 | 5945 | 5.62 | −0.30 | 13.60 |
| 0.1 | 6065 | 5.64 | −0.21 | 13.72 |
| 0.2 | 6183 | 5.65 | −0.12 | 13.83 |
| 0.3 | 6301 | 5.67 | −0.04 | 13.94 |
| 0.5 | 6545 | 5.69 | +0.14 | 14.16 |
| 0.7 | 6792 | 5.72 | +0.31 | 14.37 |
| 1.0 | 7180 | 5.76 | +0.56 | 14.58 |

$$T_{\text{eff}} = 10\,000 \text{ K}$$
$$\log g = 4.0, \log Z/Z_{\odot} = 0$$

| $\log \bar{\tau}$ | $T$ (K) | $\log P_g$ (dyn cm$^{-2}$) | $\log \bar{\kappa}_R$ (cm$^2$ g$^{-1}$) | $\log n_e$ (cm$^{-3}$) |
|---|---|---|---|---|
| $-3.0$ | 7495 | 1.54 | $-0.38$ | 12.68 |
| $-2.5$ | 7673 | 1.87 | $-0.20$ | 12.97 |
| $-2.0$ | 7875 | 2.20 | $+0.02$ | 13.26 |
| $-1.5$ | 8170 | 2.49 | $+0.27$ | 13.56 |
| $-1.0$ | 8540 | 2.74 | $+0.59$ | 13.87 |
| $-0.9$ | 8670 | 2.79 | $+0.68$ | 13.94 |
| $-0.8$ | 8810 | 2.83 | $+0.77$ | 14.02 |
| $-0.7$ | 8990 | 2.87 | $+0.87$ | 14.10 |
| $-0.6$ | 9175 | 2.90 | $+0.95$ | 14.18 |
| $-0.5$ | 9380 | 2.93 | $+1.04$ | 14.25 |
| $-0.4$ | 9600 | 2.96 | $+1.12$ | 14.31 |
| $-0.3$ | 9840 | 2.99 | $+1.21$ | 14.37 |
| $-0.2$ | 10110 | 3.03 | $+1.28$ | 14.43 |
| $-0.1$ | 10430 | 3.06 | $+1.33$ | 14.49 |
| 0 | 10790 | 3.09 | $+1.35$ | 14.54 |
| 0.1 | 11160 | 3.13 | $+1.36$ | 14.58 |
| 0.2 | 11610 | 3.17 | $+1.34$ | 14.61 |
| 0.3 | 12070 | 3.22 | $+1.32$ | 14.66 |
| 0.5 | 13250 | 3.35 | $+1.23$ | 14.76 |
| 0.7 | 14530 | 3.53 | $+1.17$ | 14.90 |
| 1.0 | 17040 | 3.87 | $+1.12$ | 15.18 |

$$T_{\text{eff}} = 10\,000 \text{ K}$$
$$\log g = 2.5, \log Z/Z_{\odot} = 0$$

| $\log \bar{\tau}$ | $T$ (K) | $\log P_g$ (dyn cm$^{-2}$) | $\log \bar{\kappa}_R$ (cm$^2$ g$^{-1}$) | $\log n_e$ (cm$^{-3}$) |
|---|---|---|---|---|
| $-3.0$ | 7289 | 0.07 | $-0.50$ | 11.61 |
| $-2.5$ | 7531 | 0.46 | $-0.39$ | 11.98 |
| $-2.0$ | 7791 | 0.85 | $-0.24$ | 12.34 |
| $-1.5$ | 8089 | 1.20 | $-0.04$ | 12.68 |
| $-1.0$ | 8540 | 1.49 | $+0.21$ | 12.99 |
| $-0.9$ | 8680 | 1.55 | $+0.26$ | 13.05 |
| $-0.8$ | 8820 | 1.60 | $+0.31$ | 13.11 |
| $-0.7$ | 8980 | 1.65 | $+0.36$ | 13.16 |
| $-0.6$ | 9160 | 1.70 | $+0.40$ | 13.22 |
| $-0.5$ | 9370 | 1.75 | $+0.44$ | 13.28 |
| $-0.4$ | 9590 | 1.81 | $+0.47$ | 13.33 |
| $-0.3$ | 9840 | 1.86 | $+0.49$ | 13.38 |
| $-0.2$ | 10140 | 1.92 | $+0.50$ | 13.43 |
| $-0.1$ | 10460 | 1.98 | $+0.51$ | 13.48 |
| 0 | 10810 | 2.05 | $+0.51$ | 13.54 |
| 0.1 | 11180 | 2.12 | $+0.50$ | 13.60 |
| 0.2 | 11620 | 2.20 | $+0.49$ | 13.67 |
| 0.3 | 12110 | 2.29 | $+0.48$ | 13.73 |
| 0.5 | 13220 | 2.48 | $+0.45$ | 13.88 |
| 0.7 | 14540 | 2.68 | $+0.42$ | 14.05 |
| 1.0 | 17000 | 2.99 | $+0.41$ | 14.32 |

$$T_{\text{eff}} = 30\,000 \text{ K}$$
$$\log g = 2.5,\ \log Z/Z_{\odot} = 0$$

| $\log \bar{\tau}$ | $T$ (K) | $\log P_g$ (dyn cm$^{-2}$) | $\log \bar{\kappa}_R$ (cm$^2$ g$^{-1}$) | $\log n_e$ (cm$^{-3}$) |
|---|---|---|---|---|
| −3.0 | 19676 | 1.33 | −0.36 | 12.59 |
| −2.5 | 20492 | 1.78 | −0.29 | 13.03 |
| −2.0 | 21455 | 2.20 | −0.21 | 13.43 |
| −1.5 | 22990 | 2.61 | −0.11 | 13.81 |
| −1.0 | 24900 | 3.01 | −0.01 | 14.17 |
| −0.9 | 25430 | 3.08 | +0.01 | 14.23 |
| −0.8 | 26040 | 3.16 | +0.03 | 14.30 |
| −0.7 | 26730 | 3.24 | +0.05 | 14.37 |
| −0.6 | 27460 | 3.31 | +0.07 | 14.44 |
| −0.5 | 28280 | 3.39 | +0.09 | 14.50 |
| −0.4 | 29170 | 3.47 | +0.11 | 14.56 |
| −0.3 | 30130 | 3.54 | +0.13 | 14.62 |
| −0.2 | 31170 | 3.62 | +0.15 | 14.69 |
| −0.1 | 32300 | 3.70 | +0.18 | 14.75 |
| 0 | 33500 | 3.77 | +0.20 | 14.80 |
| 0.1 | 34790 | 3.84 | +0.24 | 14.86 |
| 0.2 | 36160 | 3.91 | +0.27 | 14.92 |
| 0.3 | 37660 | 3.98 | +0.30 | 14.98 |
| 0.5 | 40920 | 4.13 | +0.33 | 15.09 |
| 0.7 | 44550 | 4.29 | +0.33 | 15.22 |
| 1.0 | 51350 | 4.56 | +0.29 | 15.43 |

[a] Interpolated from Kurucz 1979, *Ap. J. Suppl.*, **40**, 1.

# References

Abell, G. O., 1982, *Exploration of the Universe*, Fourth Edition, CBS College Publishing.

Auer, L. and D. Mihalas, 1973, *Astrophysical Journal Supplement*, **24**, 193.

Bhatnagar, P. L., M. Krook, D. H. Menzel, and K. N. Thomas, 1955, in *Vistas of Astronomy*, ed. A. Beer, Vol. 1, 296.

Biermann, L., 1951, *Zeitschrift für Astrophysik*, **29**, 274.

Böhm-Vitense, E. and T. Dettmann, 1980, *Astrophysical Journal*, **236**, 560.

Böhm-Vitense, E., 1979, *Mercury*, **8**, 29.

Böhm-Vitense, E., 1986, *Astrophysical Journal*, **301**, 297.

Chalonge, D. and V. Kourganoff, 1946, *Annales d'Astrophysique*, **9**, 69.

Chapman, S., 1957, *Smithsonian Contribution Astrophysics*, **2**, 1.

Crawford, D. L., 1958, *Astrophysical Journal*, **128**, 185.

Davidson, C. R. and F. J. M. Stratton, 1927, *Memoirs Royal Astronomical Society*, **64**, IV.

Deutsch, A., 1956, *Astrophysical Journal*, **123**, 210.

Edlén, B., 1941, *Arkiv for Matematik Astronomi och Fysik*, **28**, B, No. 1.

Edlén, B., 1942, *Zeitschrift für Astrophysik*, **22**, 30.

Grotrian, W., 1939, *Naturwissenschaften*, **27**, 214.

Hartmann, L., A. K. Dupree, and J. C. Raymond, 1980, *Astrophysical Journal (Letters)*, **236**, L143.

Hartmann, L. and K. B. McGregor, 1980, *Astrophysical Journal*, **242**, 260.

Hundhausen, A. J., 1972, *Coronal Expansion and Solar Wind*, Springer Verlag, New York, Heidelberg, Berlin.

Inglis, D. R. and E. Teller, 1939, *Astrophysical Journal*, **90**, 439.

Johnson, H. L., 1965, *Astrophysical Journal*, **141**, 940.

Jordan, C., A. Brown, F. M. Walter, and J. L. Linsky, 1985, *Monthly Notices Royal Astronomical Society London*, **218**, 465.

Keenan, P. C. and R. E. Pitts, 1980, *Astrophysical Journal Supplement*, **42**, 541.

Kuiper, G. P., 1953, *The Sun*, University of Chicago Press, Chicago, Illinois.

Lighthill, M., 1952, *Proceedings Royal Society London Ser. A*, **211**, 564.

Linsky, J. L. and B. M. Haisch, 1979, *Astrophysical Journal (Letters)*, **229**, L27.

Moore, C. E., 1959, *A Multiplet Table of Astrophysical Interest*, NBS Technical Note No. 26, United States Department of Commerce.

Morgan, W. W., H. A. Abt, and J. W. Tapscott, 1978, *Revised MK Spectral Atlas for Stars Earlier than the Sun*, Yerkes Observatory, University of Chicago and Kitt Peak National Observatory.

Morgan, W. W., P. Keenan, and E. Kellman, 1943, *An Atlas of Stellar Spectra*, University of Chicago Press, Chicago, Illinois.

Parker, E. N., 1958, *Astrophysical Journal*, **128**, 664.

Proudman, I., 1952, *Proceedings Royal Society London Ser. A*, **214**, 119.

Rosner, R., W. H. Tucker, and G. S. Vaiana, 1978, *Astrophysical Journal*, **220**, 643.

Schmidt-Kaler, T., 1982, in *Landolt-Börnstein: Numerical Data and Functional Relationships in Science and Technology, New Series, Volume 2, Astronomy and Astrophysics*, ed. K. Schaifers and H. H. Voigt, Springer Verlag, Berlin, Heidelberg, New York, p. 1.

Skumanich, A., 1972, *Astrophysical Journal*, **171**, 565.

Stencel, R., D. J. Mullan, J. L. Linsky, G. S. Basri, and S. P. Worden, 1980, *Astrophysical Journal Supplement*, **44**, 383.

Strömgren, B., 1963, in *Basic Astronomical Data*, p. 123, ed. K. Strand, Vol. III of *Stars and Stellar Systems*, gen. ed. G. P. Kuiper and B. M. Middlehurst, University of Chicago Press.

Unsöld, A., 1977, *The New Cosmos*, Springer Verlag, New York, Heidelberg, Berlin.

Vidal, C., J. Cooper, and E. Smith, 1973, *Astrophysical Journal Supplement*, **25**, 37.

Wildt, R., 1939, *Astrophysical Journal*, **90**, 611.

Zirin, H., 1966, *The Solar Atmosphere*, Blaisdell Publishing, Waltham, Massachusetts.

### Other books dealing with stellar atmospheres

Aller, L., 1963, *The Atmospheres of the Sun and the Stars*, Second Edition, Ronald Press Company, New York.

Chandrasekhar, S., 1950, *Radiative Transfer*, First Edition, Oxford University Press; 1960, Second Edition, Dover Publications, Inc., New York.

Gibson, E. G., 1972, NASA Spec. Publ. SP-303.

Gingerich, O., editor, 1969, *Theory and Observations of Normal Stellar Atmospheres*, MIT Press, Cambridge, MA.

Gray, D., 1976, *Observations and Analysis of Stellar Photospheres*, John Wiley and Sons, Inc., New York.

Greenstein, J., editor, 1960, *Stellar Atmosphere*, Vol. VI of *Stars and Stellar Systems*, gen. ed. G. P. Kuiper and B. M. Middlehurst, University of Chicago Press, Chicago, Illinois.

Jefferies, J. T., 1968, *Spectral Line Formation*, Blaisdell Publications, Waltham, MA.

Kourganoff, V., 1952, *Basic Methods in Transfer Problems*, First Edition, Oxford University Press; 1963, Second Edition, Dover Publications, Inc., New York.

Mihalas, D., 1978, *Stellar Atmospheres*, Second Edition, Freeman and Company, San Francisco.

Novotny, E., 1973, *Introduction to Stellar Atmospheres and Interiors*, Oxford University Press, New York.

Swihart, T. L., 1981, *Radiative Transfer and Stellar Atmospheres*, Pachart Publishing House, Tucson.

Swihart, T., 1968, *Astrophysics and Stellar Astronomy*, John Wiley and Sons, Inc., New York.

Swihart, T., 1971, *Basic Physics of Stellar Atmospheres*, Pachart Publishing House, Tucson.

Unsöld, A., and Baschek, B., 1982, *The New Cosmos*, Springer Verlag, Heidelberg, Berlin, New York.

Unsöld, A., 1955, *Physik der Sternatmosphären*, Springer Verlag, Berlin, Göttingen, Heidelberg.

# Index